OPEN YOUR HEART TO GOD
THROUGH BHAKTI YOGA:
YOGA OF DEVOTION

This is a comprehensive text on Bhakti Yoga (Yoga of Devotion). It covers all aspects of Bhakti Yoga with detailed guidelines for its practice. It is one of the four major paths for attaining the life's goal of Self-realization or God-realization. Compared with the other paths, the practice of Bhakti Yoga is far simpler, less austere and more natural. Everyone may not have exceptional energy to practise Karma Yoga, or keen intellect for practising Jnana Yoga, or one-pointed concentration for practising Raja Yoga, but all have love in their hearts for practising Bhakti. Therefore as the Divine Sage Narada states, Bhakti is the easiest of the spiritual paths.

Though all the paths of Yoga lead to the same goal, Bhakti is considered to be superior to the other paths in some respects. In all other paths, there is the fear of a fall, but in Bhakti there is no such fear. The other paths are methods of practice and are needed as long as one has not realized the Supreme. They only help the aspirants on their way to the goal. But supreme devotion (Para Bhakti) is, on the other hand, the goal in itself. Bhakti is both the means and the end. Supreme Devotion and God Realization are identical. Bhakti is a living force in the spiritual life of the world. Throughout the ages many devotees have realized God by deep devotion and self-surrender.

OPEN YOUR HEART TO GOD THROUGH BHAKTI YOGA: YOGA OF DEVOTION

Based on the Teachings of
**Sri Swami Satchidananda and
Other Great Masters**

Dr. O.R. Krishnaswami

Integral Yoga Centre
Gibraltar

DEV BOOKS
Delhi

Published for

Integral Yoga Centre
33 Town Range
Gibraltar

ISBN 81-89835-01-7
First published 2006
© 2006, **Krishnaswami**, O.R.

Typeset, printed and published by
DEV BOOKS
11-B Court Road,
Delhi-110054.
email:devbooks@vsnl.net

Contents

Preface

This Text on Bhakti Yoga (Yoga of Devotion) was assembled from the transcribed lectures of our Rev. Sri Gurudev Sri Swami Satchidananda, and from the *Narada Bhakti Sutras* and their commentaries, the teachings of Bhagavan Sri Ramakrishna, and the works of Swami Vivekananda, our Grand Master Swami Sivananda, Swami Jyotirmayananda, Swami Yatiswarananda and other Saints, which are listed under *References*. I acknowledge my grateful indebtedness to all of them. If there are any errors and faults, I am personally responsible.

This Text is a comprehensive one, covering all aspects of Bhakti Yoga with detailed guidelines for practising the great Yoga of Divine Love. There are four major paths for attaining the life's goal of Self-realization or God-realization, namely Bhakti Yoga, Karma Yoga, Jnana Yoga and Raja Yoga. Karma Yoga is the approach to God through selfless service. Jnana Yoga is the Path of Wisdom. It is the approach through dispassion and discrimination between the real and unreal. Raja Yoga is the approach through intensive practice of meditation. These three Yogas —Karma Yoga, Jnana Yoga and Raja Yoga—demand qualities and faculties that are not possessed by every one or even by a large majority of human beings. Karma Yoga calls for exceptional energy and strong will as well as great humility and patience. Jnana Yoga requires an exceptionally acute intellect and reason. Raja Yoga calls for unwavering concentration and control of senses.

Compared with them, tne practice of Bhakti Yoga seems far simpler, less austere and more natural. Everyone may not have exceptional energy, keen intellect or concentration, but all have love in their hearts. Therefore as the Divine Sage Narada states, Bhakti is the easiest of the Yogas. The very fact that very ordinary persons have become great devotees and realized God is a valid proof for this statement.

The compilation of this work was entirely due to the Grace of our beloved Sri Gurudev and the Supreme Lord. Indeed Sri Gurudev and the Supreme Lord worked through this humble soul. With deep Devotion I dedicate this Text to the Lotus Feet of Sri Gurudev and the Lord for having used this soul as an instrument in their Hands.

May this Text inspire and guide the students of Yoga and Devotees in their sincere practices for achieving the ultimate purpose of life, God-Realization.

O. R. KRISHNASWAMI

December 25, 2005

Acknowledgments

I am deeply grateful to Sri Swami Satchidananda Maharaj whose teachings inspired me to compile this text as an instrument in his hands.

I am profoundly grateful to Rev. Swami Karunananda for having given a beautiful foreword to this book.

I am indebted to the authors of books under references and cited in footnotes, from whose thoughts I derived enlightenment.

I am thankful to Swami Murugananda who archived all of Gurudev Sri Swami Satchidananda's satsangs on his computerized database of the Sastri Project, from which references to Sri Gurudev's teachings were taken.

I am also thankful to my sons and daughter who provided me with a laptop computer and also procured many books for my study and reference.

I am grateful to the Governing Council of Satchidananda Ashram, Yogaville, Buckingham, Virginia for having provided me with all the facilities and support for working on this project.

The publication of this text is entirely due to the dedicated efforts of Mrs. Nalanie Chellaram, a supreme devotee of Sri Gurudev Swami Satchidananda Maharaj. She mobilized the generous support of some sponsors from Integral Yoga Gibraltar and made the arrangements for the publication of this work. Words are inadequate to convey my sincere thanks to beloved sister Nalanie and the benevolent sponsors who preferred to remain anonymous.

Also to Paddy and Julio Alcantara and Hersha Chellaram for their help with the editing.

Above all I offer my deep prayers to God by whose Grace all these have become possible.

O.R. KRISHNASWAMI

December 25, 2005

1
The Goal of Life

Basic Needs and Desires

We human beings, like other living beings, have some basic needs–
need for food, shelter, self-preservation and so forth. In addition,
we are attracted to various objects and experiences, which satisfy
our sense of ego. Desires are endless. As a result of scientific
discoveries and technological inventions, innumerable things have
been created for satisfying our ever-growing desires. Impelled by
our urges and desires we earn money through various means and
acquire various material objects and comforts. Many of us have no
hesitation to resort to unfair means for acquiring wealth. We aspire
for position, power, control over others, name and fame, and are
engaged in endless activities. What is the underlying purpose behind
such desires, aspirations and the activities? The underlying purpose
is to be happy.

Happiness

Can we find real happiness? Where do we look for happiness? We
look for happiness from outside sources, and chase the fleeting
shadows of material wealth, possessions, positions, power and
prestige. We ransack a thousand and one ways for attaining
happiness. We compete with one another in this mad race. Successful
and lucky people become millionaires; unsuccessful and unlucky
people become paupers. While the poor plunge in squalor and

poverty, the rich float in affluence with all kinds of luxuries. But do the rich really have peace and happiness? No. They have greater worries, anxieties and stress than anybody else. They are restless without real happiness. This is like a group of people racing around in a circle. One person overtakes another, and sometimes, the other overtakes him. Everyone runs fast and competes to overtake but all of them remain in the same place. Life in the present time is like that race. It is becoming faster and faster due to possessive tendencies and greed, but still it is encircled around the objects of the five senses.

Material wealth, no doubt, is necessary for a decent healthy living, but it should come through honest means. Rich people should hold their surplus wealth as trustees, as stressed by John Ruskin[1] and Mahatma Gandhi,[2] and utilize it for the benefit of the needy people. Greed and extravagance are not going to bring any happiness and peace.

Why do we run after material wealth, position, power and sensual pleasures? We all seem to be thinking that happiness is something that comes from the outside, from material objects and sensual pleasures, something to be grabbed from somewhere. The happiness that comes from outside is momentary and it is followed by manifold miseries. Every external object that gives us some fleeting happiness is subject to change. It will disappear one day. Beauty, power, position, wealth, relationships, name, and fame—all such things come and go. None of these things are permanent. So they cannot give us any lasting happiness.

The Nature of Material Happiness

What is the true nature of material happiness? Simply wanting to do something or get something to be happy. The very desire makes you unhappy until you get it fulfilled. But before the desire arose you were peaceful. Now the desire prompts you to make all the efforts to get it satisfied. When you get the object or experience you wanted, the desire is fulfilled and the mind feels peaceful again. You may feel happy that your desire was fulfilled. You think that the object or experience brought the happiness, but what really happened was you just regained the mental tranquillity that was there before the desire arose. The object or experience just satisfied your want, which disturbed your happiness in the first place.

Now you got what you wanted, you feel happy; but immediately the fear of losing it comes in. You start worrying about it. So no permanent happiness can come from merely satisfying desires.

Permanent Happiness

Is there no permanent happiness? Yes, there is, but it does not come from outside. It exists within our heart. Our very real nature itself is peace and joy. We are not aware of that, and go on searching for it outside, without proper direction. This can be illustrated in a story of an old woman who was sewing her clothes.

She dropped her needle, so she immediately began to look for it by the light of a street lamp. A friend of hers was passing by and asked her what she was doing.

> She said, "I am looking for the needle."
> An hour passed by, but neither of them could find it. Finally the friend asked her, "Where did you lose it?"
> "In the house," she said.
> "Then why are we looking for it outside?"
> "Because there is no light inside; the only light is out here."

We are no different from that old woman. We look for happiness outside. By searching for it from material things and experiences, we continue to disturb our inner peace and joy by getting entangled in a continuous stream of desires and the consequent anxieties, worries, stress, disappointments, distress and dejection. Why do we act like this? It is because of our ignorance of our real nature.

Ignorance of our Real Nature

What is our real nature? Are we the body-mind-senses complexes? No. We are not the bodies, nor the minds, nor the senses. They are only our external covers. We are the eternal Self, or Atman, or Spirit within. The Self is the God within us. The Self has neither form nor shape. It is Consciousness or Awareness. It is eternal. It has no beginning, no end. It is omnipresent, omnipotent and omniscient. Mundaka Upanishad[3] declares:

"This Self is effulgent, all-pervading, beyond birth,
Without the vital forces, and devoid of mind." (2. 1. 2)

Kena Upanishad says:

"The Self or Atman is beyond all mental concepts,　Beyond all names and senses." (2. 1)

Katha Upanishad reveals:

"The Atman (Self) though hidden in all beings, does not shine forth,
But can be seen by the Seers through their sharp and subtle intuition."

(1.3.2)

Bhagavad Gita[4] says:

"The Self is not born, nor does it ever die…. Unborn, eternal,
Changeless and ancient, it is not killed, when the body is killed." (2. 20)
"This Self cannot be cut, burnt, wetted, or dried up.
It is eternal, all-pervading, stable, ancient and immovable." (2. 24)

Our real being is the infinite *Atman* or Self. This unchanging and eternal *Atman* is the real nature of everyone. Why then, most of us are not aware of our real nature? Out of ignorance, we falsely identify ourselves with our body-mind complexes. The natural tendency of our mind is to run outward towards the objects of senses. The senses are drawn naturally to external objects, and the mind is attached to the senses because of ignorance. From ignorance arises the ego-sense. Then comes attachment to pleasurable things of the world, aversion to unpleasant things, and finally thirst for life. All these lead to restless activities for the purpose of satisfying our endless desires. As wants are insatiable, we are always discontent and dissatisfied.

Man has made unprecedented progress in getting the knowledge of the external world. He has invented unimaginable devices, gadgets and equipment. He has even learnt to harness atomic energy, make powerful destructive devices, and to land on the moon. Despite this progress in his outer world, there is not much progress in his inner world. He is engrossed in ignorance. He is restless and blindly clings to earthly existence. He is perverted, self-centred and selfish. He is ready to do anything to earn money and satisfy his egoism. He does not hesitate to produce and sell toxic materials like cigarettes, liquors, and meat, which cause deadly diseases to his fellow people. He sells powerful destructive weapons to countries that fight with one another. Human beings, at times, become so fanatic as to kill their own fellow persons in the name of religion and ethnic difference. What is the outcome? The world is plagued with miseries, diseases, violence, strife and discard. "Even if a small fractional part of the human beings," says Swami Vivekananda (1863–1902) [5], "can put aside the idea of selfishness, narrowness and littleness, this earth will become a paradise tomorrow."

Know Thyself

What is the way out? "Know thyself" is the answer; but how can we know ourselves? Is it mere intellectual knowledge that we need? Modern psychology may explain many functions of the mind and make interesting guesses; but it is a study of the mind as an object. That does not help us to know ourselves. The ancient realized sages and seers of India who have completely mastered the secrets of life tell us that for the pursuit of true happiness one must take his thoughts and vision away from the vanishing objects of the external world, and instead, turn inward and go deep into the recesses of his own mind. There he will find the immortal Self, the very source of self-existent and ever-lasting peace and happiness. He, who has discovered this mysterious source, looks upon the material world with wonder, coupled with pity. The Chandogya Upanishad, composed several thousand years ago, declares:

"That which is the subtle essence, the whole universe has for its Self; That is the Truth; That is the Atman or Self; and That thou art." (6. 8. 7)

"The Self, which is free from evil, free from old age, free from death,
Free from grief... should be sought and understood.
One who has found out and who understands that Self has
Obtained all worlds and achieved all desires." (8. 7. 1)

When that higher knowledge – enlightenment – dawns, it destroys our
ignorance, just as sunrise removes darkness. Then we experience our own
true Self, the Atman within. This knowledge of Self is pure knowledge. In
Bhagavad Gita, Lord Krishna, God-incarnate, says:
"When one sees the One indestructible Reality (Self) in all beings
(or Infinity in finite things), then one has pure Knowledge." (18. 20)

The attainment of this knowledge–the Knowledge of Self – is the
ultimate purpose of life. Only a human being can attain this
knowledge. How is this supreme knowledge attained? The answer
of the Upanishads is quite definite. Atman, the mystery of our life,
the light of our soul, is something beyond reason, and therefore, it
can never be attained by reason alone. Katha Upanishad declares:

"Not through instruction is the Self attained, not by the intellectual power,
Not even through much hearing. It is to be attained only by one whom the
Self chooses. To such a one the Self reveals Its own nature." (I. 2. 23)

"Not through deep knowledge can the Atman be reached, unless
evil ways are abandoned and tranquility exists, mind is concentrated and
composed." (I. 2. 24)

"Know the Self as the Lord of a chariot and the body as the chariot.
Know that intellect as the charioteer, and the mind as reins." (I. 3. 3)

"The senses are the horses; the objects of senses the paths; (the soul)
Associated with the body, the senses and the mind is the enjoyer." (I. 3. 4)

"He, whose chariot is driven by reason, and who controls the reins of his mind, reaches the end of the Journey, that Supreme Abode of the All-pervading."(I.3.9)

"The Self, though hidden in all beings, is invisible, but can be
Seen by the Seers through their sharp and subtle intuition." (I. 3. 12)

This is described as Self-realization or God-realization. This is the goal of life. The Upanishads declare:

"Blessed is the one who realizes God in this life; if not he has lived in vain."
Sankaracharya (788 – 820)⁶ says, "A man is born not to desire life in the world of the senses, but to realize the bliss of a free soul in union with God."

Yoga⁷ is the means through which this goal is reached. It is self-training for Self-realization, which is a state of eternal peace and joy. One, who attains Self-realization, sees that the Self or God in himself is the same God in all beings. Ishavasya Upanishad declares that the Sage sees all beings in the Self, and the Self in all beings. He crosses the ocean of life. (6) He is liberated from bondage, and enjoys eternal peace and bliss. (For details, *see* chapter 30. God-realisation.)

References

1. A British writer and critic, and a Social Thinker. His book entitled *"Unto The Last"* transformed Mahatma Gandhi's life into a great social reformer. The teachings of this book formed the basis Of Gandhi's philosophy of *Sarvodaya* (Welfare of All).
2. The Father of India, who obtained the freedom of India from the British Rule through bloodless revolution of Satyagragha; a Saint in the political field.
3. The *Upanishad*s are the ancient scriptures of India. They are the concluding portions of the *Vedas*, which are the foundational Scriptures of the Hindus, revealed by God to the great ancient Seers (*Rishies*) of India. The *Upanishad*s contain the essence of the *Vedas*. They are the source of the Vedanta philosophy and the divine intuitive knowledge.

Manduka Upanishad is one of the Oldest *Upanishad*s.

4. An ancient Hindu scripture containing the essence of *Upanishads* and Yoga.

5. The greatest disciple of Sri Ramakrishna, the God-man of modern India; founded the Ramakrishna Mission and introduced the Vedanta Philosophy to the West. (For a brief biography of Sri Vivekananda, *see* the chapter 33. World Teachers and Great devotees.)

6. One of the greatest Saint and philosophers of India, an exponent of Advaita (Non-dualistic) Vedanta. He spread this philosophy throughout India and established ashrams in the Four Corners of India.

7. For details *see* the next chapter.

2

Yoga and the Goal of Life

Yoga is the path, traversing which a human being ultimately attains Self-realization or God-realization, the ultimate goal of life. The Sanskrit word *Yoga* was derived from its root "*Yuj*" which means "to join" or "to unite." What are the two things which join together? The human soul or *jivatman* unites with the Supreme Spirit, the Divine Reality or *Paramatman*. Although the *jivatman* is a facet of the *Paramatman*, (and in essence both are the same) the *jivatman* has become subjectively separated from the *Paramatman*. However, after going through an evolutionary cycle in the manifested universe, it experiences its true nature. This is described as reunion with the *Paramatman*. This apparent union of a *jivatman* with the *Paramatman*, and the method through which this union is attained, both are called Yoga. Just as there are several paths to the summit of a mountain, there are several paths to the realization of the indwelling Divinity or Self. They primarily consist of the Path of Devotion (*Bhakti Yoga*), the Path of Selfless Action (*Karma Yoga*), the Path of Knowledge (*Jnana Yoga*) and the Path of Meditation (*Raja Yoga*). The word "*Yoga*" in a generic sense refers to all the different paths or methods used to obtain this union. The reason for the existence of several paths is to suit the varying temperaments and capacities of human beings, but all the paths lead to the same goal of Self-realization or liberation.

At the start one may be described as following a particular path. But with spiritual growth all the paths become more or less harmonized in one's spiritual life. This is the great lesson we learn from the *Bhagavad Gita*, which is the very essence of all Hindu scriptures.

All aspirants following the different paths attain the same supreme peace and bliss. Sri Krishna points this out distinctly:

"Take refuge in the Divine with all thy being. By His grace
You shall attain supreme peace and the eternal abode." (*Gita*, 18. 62)

"The well-poised, forsaking the fruits of action,
Attains peace born of steadfastness…" (*Gita*, 5. 12)

"Having attained knowledge, one gains supreme peace at once." (*Gita*, 4.39)

"Always keeping the mind steadfast in Yoga, the Yogi of subdued mind
Attains to peace that abides in Me (Lord) and

Culminates in Liberation." (*Gita*, 6.15)

The *Bhagavad Gita* places different paths of Yoga before aspirants of different temperaments and tendencies. They are discussed here in brief.

Bhakti Yoga

Bhakti Yoga is a Yoga of Devotion, a path of love for God. It is an intense longing and love for God. It enables an aspirant to constantly remember God. It purifies his emotions and elevates his mind to the consciousness of the Reality. He surrenders all his thoughts, words and deeds to the Lord and adores Him with an unflinching devotion. By constant meditation on the Lord, the devotee imbibes into his own being the Divine attributes.

The devotee yearns for union with his beloved Lord with all his heart. Worldly love is not able to satisfy this deep passion of the soul. His entire mind flows, as it were, in a continuous stream towards his Lord who is of the nature of supreme love. His God is not a being dwelling somewhere in a far off heaven, but is seated in his own heart. In order to see Him and to feel His Divine presence he has only to look within. In the *Bhagavad Gita* Sri Krishna says:

"The Lord dwells in the heart of all beings…. Take refuge in Him
With all your being. By His grace you shall attain supreme peace
And the eternal abode." (18. 61-62)

Karma Yoga

Karma means action. So *Karma Yoga* means Yoga of Action. The tendency to perform action is innate in all beings, and none can remain inactive. Of those who are particularly of an active temperament, what is required is not cessation from action but the proper controlling and directing of their activity. Work with attachment to its fruit binds the soul, but selfless work performed in a spirit of worship becomes a means of attaining freedom. Selfless action does not bind the doer, as he has no sense of doership or attachment to the fruits of his action. Says Sri Krishna in the *Bhagavad Gita*:

"One who does action, forsaking attachment and offering to the
Divine, is not tainted by evil as a lotus leaf by water." (5. 10)

"Though ever engaged in actions, he having taken refuge in Me (the Lord),
Attains to the eternal and immutable state by My grace." (18. 56)

The work that is offered to God becomes worship and Yoga. This Yoga helps a person to live peacefully and usefully in the world while remaining above it, unaffected by its fetters, like a lotus leave on the water. Selfless action purifies his mind and ultimately leads him to Self-realization.

Jnana Yoga

Jnana means knowledge. So *Jnana Yoga* is Yoga of Knowledge. This path involves intense discrimination (*Viveka*). One with a contemplative mind tries to seek answer to a basic question: "Who am I? Am I a body or mind? If not, what else?" The aspirant discriminates between the real and the unreal, between the eternal and transient. Such persistent discrimination and self-inquiry ultimately leads to Enlightenment or the Knowledge of Self. Ramana Maharishi, the great Indian sage of the twentieth century, was a great *Jnana Yogi.*

The *Jnani*, or the seeker of knowledge, with a philosophical turn of mind, is not satisfied with the little pleasures of this life. He is anxious to know his true nature and the reality at the back of the phenomenal world. He does not feel any great attraction for the personal aspects of the Divinity. He yearns to attain to his true Self. Speaking of those who follow the path of knowledge, Sri Krishna observes in the *Bhagavad Gita*:

"Those, who have their intellect absorbed in That, whose Self is That,
Established in That as their supreme goal, attain to the highest salvation,
Their impurities being cleansed by knowledge." (5. 17)

"Relative existence has been conquered by them, even here in this world,
Whose mind rests in evenness. Since the infinite Self is even and without
imperfection, therefore they are established in Him." (5. 19)

Raja Yoga

This "Royal Path" is highly scientific. It is as old as human civilization. It is the Yoga of Meditation. This ancient Yoga was codified by the saint, Patanjali, in his *Yoga Sutras*. This Yoga concerns itself with three realms—the physical, the mental and the spiritual. It is an eight-step path consisting of Restraints (*Yamas*), Observances (*Niyamas*), Postures (*Asanas*), Regulation of Life force (*Pranayama*), Withdrawal of the Senses (*Pratyahara*), Concentration (*Dharana*), Meditation (*Dhyana*), and Absorption (*Samadhi*). By practicing *Raja Yoga* regularly a seeker learns to control his desires, emotions and thoughts and purifies his mind, and finally attains *Samadhi*, the state of Super-consciousness.
The person of a contemplative bent of mind with tremendous

willpower wants to control the whole of Nature, external and internal. But sporadic attempts at concentration will lead only to failure and mental depression. The restless mind is to be brought under control slowly and steadily by regular practice and renunciation. Sri Krishna says:

"With steady understanding and fixing the mind on the Self,
Let the Yogi attain quietude by degrees. By whatever cause
The restless and unsteady mind wanders away,
Let him, curbing it from wandering, bring it back
Under the control of the Self alone.... Verily supreme bliss
Comes to the Yogi whose mind has become perfectly tranquil, whose
Passions are quieted, who is free from impurities,
And who has become one with the all-pervading Being." (*Gita,* 6. 25-27)

Complementarities of the Paths

The different paths of Yoga are not mutually exclusive. They are complementary to one another. One cannot really follow any one of them to the exclusion of the rest. An aspirant may only lay greater emphasis upon one or the other. For instance, meditation is to be practiced as a discipline, no matter which path the aspirant may follow. Besides, he must discriminate and acquire spiritual knowledge to practice even *Bhakti Yoga* and *Karma Yoga*. And every aspirant, whatever may be the Yoga that he follows, is to be active without expecting the fruits of actions or dedicating them to the Lord. Above all, every aspirant, irrespective of the path chosen by him, must have devotion to God. Thus there must be a combination of Yogas in every aspirant's life, because perfection can be attained only by developing all aspects of one's own personality – physical, mental, moral and emotional, i.e., overall spiritual development.

One-sided development is fraught with dangers. Work sometimes becomes aimless and leads to mere restlessness. Yoga of Meditation at times degenerates into physical mortification and pursuit of psychic powers. Devotion often deteriorates into meaningless sentimentalism. And knowledge may lapse into dry intellectualism. Hence there is a great need for combining the different paths in order to safeguard against the danger of following one path to the exclusion of others. Let work be combined with meditation and self-surrender to the Lord, knowledge tempered with devotion. Let us try to be equally established in all the Yogas and bring about a harmonious development of all our latent spiritual faculties. This all-sided development of personality is the healthiest approach for any spiritual aspirant to follow.

This is why in the teachings of *Bhagavad Gita* and of Sri Ramakrishna[1] emphasis has been laid upon synthesizing all the Yogas in one's life. In the Integral Yoga Institutes established by Sri Swami Satchidananda, the Apostle of Peace, this total approach is rightly called Integral Yoga. "Integral" means a synthesis or harmonious blending. Integral Yoga is a combination of *Hatha Yoga*, *Pranayama* (control of the life force), *Bhakti Yoga*, *Karma Yoga*, *Jnana Yoga* and *Raja Yoga*. It provides for a balanced overall development of all different aspects—physical, vital, mental, moral, intellectual, intuitional and emotional aspects—of an individual's personality. Of course, everyone may lay a greater emphasis on the practice of a particular path of Yoga.

In the very life of the eternal Sage Narada, the author of the *Narada Bhakti Sutras* (*N.B.S.*), we find the integration of all the Yogas – the Paths of Devotion, Selfless Service, Wisdom and Meditation.

Narada narrates his experience:

"Let me tell of my past incarnation.... My mother was a servant in a retreat where great sages lived. I was reared in close association with them, and I too served them. While I was thus living in the society of those holy men, my heart was purified. One day, one of the sages initiated me into the sacred mysteries of wisdom. The veil of ignorance was removed from me, and I knew my real Self as Divine. Then I learnt this lesson – that the greatest remedy for all the ills of life, physical or mental, is to surrender the fruits of *Karma* (work) to the Lord. *Karma* places us in bondage, but by resigning our *Karma* to the Lord we are freed. Work that we perform as service to the Lord creates love and devotion in us. This love and devotion, in turn, brings wisdom; and at last, guided by this wisdom, we surrender ourselves to the Lord of Love and meditate upon Him. Thus I attained to wisdom and love."

In the above quotation we find the integration of all the Yogas – the Paths of Union with God through Selfless Action, Devotion, Wisdom and Meditation.

Harmony of the Yogas

All the four main paths are not, as mentioned earlier, like watertight compartments. They are all so interdependent that many of the texts speaking of one are applicable to the others as well. All the aspirants should have common virtues and also have devotion, discrimination, concentration, non-attachment, and other elements of spiritual life. Thus it is practically impossible to draw a dividing line between different types of aspirants and the paths they pursue. The following verses of the *Bhagavad Gita* illustrate the common features of the Yogi:

"Content with what comes to him without effort, unaffected by the pairs of Opposites, free from envy, and even-minded in success and failure, The wise is not bound even though he be acting." (4. 22)

"He who, being established in unity, worships Me (the Lord) dwelling in All beings – abides in Me, whatever may be his mode of living." (6. 31)

"Renouncing mentally all deeds to Me, having Me as the highest goal, and
Resorting to the Yoga of discrimination, ever fix your mind on Me." (18.57)

All these words of Sri Krishna are more or less applicable to very type of Yogi. There is a combination of all these noble virtues in the perfect ones as also in all real aspirants.

The Synthesis of Yoga

The reason for this combination is not far to seek. The human mind is a synthetic whole possessing faculties of will, feeling, and intellect or reason. The Path of Selfless Action and that of Meditation both depend primarily on the faculty of will; in the Path of Devotion feeling is the dominant factor. In the Path of Knowledge, with its discipline of analysis and discrimination, the aspirant relies mainly on intellect or reason. But all the faculties are inseparable, and so are the Yogas. They complement one another.

In the Integral Yoga system, several faculties of human beings are harmoniously developed. When the faculties of intellect, will and feeling are fully developed, the Yogi transcends these powers of the mind, and in him the innate faculty of intuition dawns and brings him directly to the vision of the supreme Self or God within. The realization of the Self is the goal of all the forms of Yoga.

Work, worship, self-inquiry, and meditation are inseparable aspects of this integrated spiritual life. Hence it is necessary that every aspirant should possess a clear conception of the Divinity, for, as the *Bhagavad Gita* says: "…he should fix his mind on God, be devoted to Him, sacrifice unto Him, bow down to Him, surrender to Him, taking Him to be the supreme goal." (9. 34)

Bhakti – The Easiest of the Yogas

Of the four major paths of Yoga described above, the three paths – *Karma Yoga*, *Jnana Yoga* and *Raja Yoga* – are very difficult, as they demand qualities and powers that are not possessed by a large majority of people. *Karma Yoga* calls for exceptional energy, strong will as well as great humility and patience. Further it is extremely difficult to perform one's duties in a spirit of detachment, without craving the result. Attachment to the result somehow enters the mind, though one may not be aware of it.

Jnana Yoga requires an exceptionally sharp intellect and reason. To follow *Jnana Yoga* in this contemporary age is also very difficult. First, our life depends entirely on food. Second, we can by no means get rid of body-consciousness, but the Knowledge of Brahman is impossible without the destruction of body-consciousness. *Raja Yoga* calls for control of senses and unwavering concentration. Compared with these three kinds of Yoga, the practice of *Bhakti Yoga* appears to be far simpler, less austere and more natural. Everyone may not have exceptional energy, keen intellect or concentration, but all have love in their hearts.

Love is something indefinable, but it is felt and experienced in one's own heart. To love something or someone is natural. Parents love their children. Children love their parents. There is love between husband and wife, and also love among friends. The attraction that we feel for one another is the attraction of God dwelling in every one of us, but we are not aware of it. Hence the Sage Narada says, "The Path of Devotion is the easiest way to attain God." (*N.B.S.* 58) The fact that very ordinary persons like the milkmaids of Brindavan, where Sri Krishna lived in his childhood days, unlettered ones like Kabir, Nanak and Tukaram, untouchables like Nanda, Ravidas, Kannappa and Tirumangaialvar and even moral wrecks like Ratnakara, Tondaradipodialvar and Narayana Bhatta became great *Bhaktas* (Devotees) and attained God is a valid proof for this statement. (For brief details about these Devotees, *see* chapter 33. World Teachers and Great Devotees.) There is a beautiful prayer by a great Sage Prahlada, "Lord, may I attain such love for you that the worldly people have for the objects of the world." Love finds its fulfilment only when we turn it towards God, whose very nature is love. God is Love; Love is God.

Love is its own proof and does not require any other. It is self-evident. It is directly felt and experienced in one's own heart. That itself is its own validity. Its nature is peace and supreme bliss. Therefore, by following the Path of Love one can realize God more easily than by following the others.

Superiority of Bhakti

Though all the paths of Yoga lead to the same goal, *Bhakti Yoga* is considered to be superior to the other paths in some respects.

Everybody is capable of love. So everyone can have love for God. In all other paths there is the fear of a fall, but in *Bhakti* no such fear exists. Therefore the Sage Narada states that *Para Bhakti*[2] (Supreme Devotion) is superior to other paths of Yoga (*Sutras* 25 – 33). The other paths of Yoga are only methods of practice and are needed only so long as one has not realized the Supreme. They only help the aspirants on their way to the goal. *Para Bhakti* is, on the other hand, the goal itself. Bhakti is the *sadhana*, the means, and the *sadhya*, the end. *Para Bhakti* and God-realization are identical.

Narada also gives another reason for the superiority of *Para Bhakti. Sadhana* or spiritual practice in other paths is self-effort, and so one must be conscious of oneself as a separate individual. So one may get over-attached with the means and forget the goal. But a supreme devotee, as an ardent lover of God, cannot remain even for a moment forgetful of the goal. He is one who has transcended the ego-consciousness. In the other paths, the seeker may develop arrogant self-assertions and self-glorification. But a devotee has less chance of developing such arrogance because of his humility and self-surrender.

Reason

All the systems of Indian Philosophy,[3] except the *Mimamsa* system declare that the knowledge of Truth is an essential condition for achieving liberation. Buddha makes Right Understanding the first of his eightfold Noble Path. Jainism includes Right Knowledge in its Triple Panacea for the cure of *samsara*, the cycle of births and deaths. Christ says, "Thou shalt know the Truth and the Truth shall make you free." All Medieval Christian theologians and mystics emphasize that steady and methodical thought must precede any spiritual practice.

The proper place for reason is provided for in spiritual practice by the insistence on reflection (*manana*). Manu, the ancient great Hindu lawgiver, says that he who knows righteousness understands the teachings of the scriptures with the help of reason and none else.

Lord Krishna, after teaching the Knowledge of Brahman to Arjuna, tells him that he should understand it critically and then adopt the teachings as seen reasonable. Jaimni, the founder of the *Mimamsa* system of philosophy in India, says, in his *Jaimni Sutras* that wherever there is a conflict in scriptures, one should resolve it by the use of reason. (1. 3. 3 and 3. 11-12) Sankaracharya emphasizes the role of the intellect in attaining the highest realization. Thus we see that all great teachers have given the highest place for reason in spiritual practice. However, reason alone is not the means of attainment.

The mind is a homogeneous entity and cannot be cut into watertight compartments. In exercising one function consciously, the others are also exercised unconsciously. Reason by itself is like a man and love is like a woman, says Sri Ramakrishna. The man can go only up to the drawing room, while the woman can enter even the inner apartments. The *Cloud of Unknowing*[4] says "By the least longing man is led to be the servant of God, not by faultless deductions of dialectics, but by the mysterious logic of the heart." Thus reason, without the aid of feeling and will, remains a blind intellectualism. Will, not assisted by knowledge and love, remains a mere aimless activity. In fact it would be impossible for each of these functions to work in isolation. How can a person experience the Highest without making the effort to know and without the love of Truth? How can one love truly without knowing the object of his love and without serving his Beloved? Thus all the powers of the mind are interdependent and always cooperate with one another. (*N.B.S.* 29)

REFERENCES

1 Sri Ramakrishna (1836 – 1886) was the God-incarnate of India. In his life he has demonstrated the unity of all Faiths. The Ramakrishna Mission established by his great disciple Swami Vivekananda propagates his teachings. (For a brief biography of Sri Ramakrishna, *see* chapter 33. World Teachers and Great Devotees.)

2. For details *see* chapter 10. Stages of Bhakti: 2. Supreme Devotion.

3. The Schools of Indian philosophies are divided into two broad groups:

orthodox and heterodox. To the first group belong the six chief philosophical systems, namely, *Nyaya, Vaisesika, Samkhya, Yoga, Vedanta* and *Mimamsa*. They are regarded as orthodox because they accept the authority of the *Vedas*. The three heterodox systems are the *Carvakas*, the Buddhism and the Jainism. These do not believe the authority of the *Vedas*. However they do believe the authority of their own spiritual leaders and founders.

4. A book of Contemplation of Medieval Period, in which a soul is united with God. It was written in fourteenth century-England. It provides a practical guidance for attaining a direct experience of God.

3

What is God?

Introduction

Ultimately, we are all seeking God, so we should know what God is. Suppose there is an apple sitting in front of you, but you do not know what an apple is. You say, "I have never seen an apple. Perhaps there is no such thing!" If you do not know what an apple is how can you know that you have one in front of you or enjoyed its taste?

Suppose I give a talk for three hours about strawberry cake. Can you taste it? No. I will be wasting my time. Instead, I have to give you the recipe. But even that is not enough. You cannot just take the recipe book, gold-gilt it, put it on the altar, wave an incense, and say, "strawberry cake, strawberry cake." You have to get the ingredients, cook them, and eat the cake.

You can get the recipe in five minutes. God gave us wonderful a recipe in the form of ten *Yama-Niyamas* given by the ancient Indian sage Patanjali in his *Yoga Sutras,* or the Ten Commandments given to Moses in the *Holy Bible.* But many of us just put the scriptures on the altar, or debate for hours on their meaning. How many have actually done the cooking and the eating?

It is easy to talk about Enlightenment, but simply hearing or reading about the Truth are not enough. You do not pay to read the menu in a restaurant. What good is anything until you have actually tasted or experienced it? God is to be experienced in the complete silence of the mind. When you experience God, you still have to use the limited mind to describe or approximate what is happening beyond the mind. That is why Buddhists do not talk about God. Lord Buddha (Sixth century B.C)[1] himself was the son of Hindu parents. Buddha must have been aware of all the calamities and chaos arising from religious conflicts. There were hundreds of sects at war with each other in the name of God. So he said, "Let me stop this nonsense. Why talk about God? Let people experience It."

Toward the end of his life, Buddha noticed Ananda, one of his senior devotees, standing nearby looking a little sad. So He asked,

"Ananda, are you sad because I am leaving this body?"

"No sir. But there is still one unanswered question that saddens me."

"What is that question?"

"Sir, you never said anything about God. Does it mean you deny God?"

"Ananda, did I ever deny God?"

"No sir, you did not. In that case, can I say that you accept God?"

"Ananda, did I ever accept God?"

He neither denied nor accepted. What does it mean? He did not want to talk about that which cannot be well described. The day man started talking about God, he created all kinds of religious quarrels. God may have made man in Its image, but men with their different egos have differed in creating God's image. That is why each one has his own God and fights about It with others. It is better not to talk but to just realize God yourself.

One person might talk for hours and hours about God, and hundreds of people might sit in front of him and listen for hours and hours, but it is all nonsense. He has only spoken about what he can conceive in his own mind, and they can only understand God based on their own minds' capacities. That is all. The true experience of God is unexplainable.

In Hindu mythology there is one form of God called *Dakshinamoorthi* who sat with four disciples in front of him. They were all learned men; they had read all the scriptures and heard all that was to be heard, but still could not realize the Truth. So they came to *Dakshinamoorthi* and requested him to explain the highest *Brahman*, the unmanifested God. He just sat there in silence. After awhile they got up, bowed down, and said, "Swami, we have understood," and went away. Only in silence can It be explained. Not in physical silence, but in real mental silence, the wisdom dawns.

Our Limitations

There is only one God who is our Lord, who is the life or Spirit within us. God is always One, but is called by different names: the *Brahman* of the *Upanishads*, the Absolute of the philosophers, Krishna or Rama, or Shiva of the Hindus, Buddha of the Buddhists, Jehovah of the Jews, Jesus Christ of the Christians, Allah of the Muslims – all these are different names of the same God. Just as water is called differently in different languages, the same God is named differently in different religions and sects. God is pleased with any name we give him. God does not care what we call Him, but rather how we feel about Him. He is interested in our hearts, not in our heads. Because of our limitations we give Him different names and forms. Out of His greatness, He lets us to do that.

It is impossible for a limited human mind to grasp the unlimited One. So the finite mind takes a little part of the infinite that it can understand and says, "This is God for me." It is similar to ten people going to the sea with ten different containers and taking some water. If one goes with a bucket, he will get a bucketful of sea. If another takes a cup, he will get a cupful of sea. If your mind is big, you have a bigger God. We limit the unlimited One to suit our limitations.

Each mind has its own concept of God. Everybody conceives of God according to his capacity, taste and temperament. The more your mind expands, the more you understand. When you rise above all your earlier imaginings of what God is, you become one with That. God is pure and peaceful. When you become pure and peaceful, you understand God by becoming Him.

On October 28, 1882, the semi-annual *Brahmo*[2] festival was celebrated at Sinthi, a village about three miles north of Calcutta, India. On invitation Sri Ramakrishna attended the festival. The

Brahmo devotees were eagerly asking questions to learn from him. One Devotee said: "Sir why are there so many different opinions about the nature of God? Some say that God has form, while others say that He is formless. Again, those who speak of God with form tell us about His different forms. Why all this controversy?"

Sri Ramakrishna replied: "A devotee thinks of God as he sees Him. In reality there is no confusion about God. God explains all this to the devotee if the devotee only realizes Him. You haven't set your foot in that direction. How can you expect to know all about God?

Listen to a story:

"Once a man entered a wood and saw a small animal on a tree. He came back and told another man that he had seen a creature of a beautiful red colour on a certain tree. The second man replied, 'When I went into the wood, I also saw that animal. But why do you call it red? It is green.' Another man who was present contradicted them both and insisted that it was yellow. Presently others arrived and contended that it was gray, violet, blue, and so forth. At last they started quarrelling among themselves. To settle the dispute they went to the tree. They saw a man sitting under it. On being asked, he replied, 'Yes, I live under this tree and I know the animal very well. All your descriptions are true. Sometimes it appears red, sometimes yellow, and other times blue, violet, gray, and so forth. It is a chameleon. And sometimes it has no colour at all. Now it has a colour, and now it has none.'

"In like manner, one who constantly thinks of God can know His real nature; he alone knows that God reveals Himself to seekers in various forms and aspects. God has attributes; then again He has none. Only a man who lives under the tree knows that the chameleon can appear in various colours, and he also knows that the animal at times has no colour at all ..... God reveals Himself in the form which His devotee loves most. His love for the devotee knows no bounds.... In the light of Vedantic reasoning *Brahman* has no attributes. The real nature of *Brahman* cannot be described. So long as your individuality is real, the world also is real, and equally real are the different forms of God and the feeling that God is a person."[3]

God made everything in His own image. God made the sky, the waters, the earth, everything. From what? Not from any other substance, but out of Himself. In other words, He Himself became the world. The world is His manifestation.

Like electric power, God steps Himself down to us from the Absolute through sound and then through form, until we can benefit from Him. God is like a static force that expresses itself through activity. All Creation is a manifestation of that static force. There is one Cosmic Essence, all pervading, all knowing, all-powerful. As such, It has no limited form and name, but can take any. This One is the cause for all the many things that we can call His creation or manifestation. Through Yoga you can experience this yourself. God is not something that can be talked about. God must be experienced. If you want to fully know and understand God, become That. If a drop wants to know the depth of the ocean, it should become the ocean.

This is why Narada, the great Divine Sage, has avoided defining God anywhere in his *Narada Bhakti Sutras,* because to define is to limit God. God, the ultimate reality, is to be experienced. When a person experiences God, he or she is unable to express Him in terms of relative experiences. In the words of Sri Ramakrishna: "When one attains *Samadhi*, then alone comes the knowledge of *Brahman*. In that realization all thoughts cease and one becomes perfectly silent. There is no power of speech left by which to express *Brahman*."

Everything is God

Everything is God. Do not look for God in heaven or on the altar. You can find Him in all the forms around you, even in a dog. Here is a story to illustrate this:

Once there was a man who prayed to God every day in the following way, "Oh, God, I really want You to come in person to have a nice sumptuous lunch with me."

Because he was always asking in this way, God appeared one day and said, "Okay, I'll come."

"God, I am so happy. When can You come? You must give me some time to prepare everything."

"Okay. I'll come on Friday."

Before He left, the man asked, "May I invite my friends?"

"Sure," said God. Then He disappeared.

The man was so excited. He invited everybody he could think of, saying, "God, Himself, is coming to my house for lunch!" Very early he started preparing all kinds of delicious dishes.

Friday at noon, everybody was there. A huge, specially decorated dining table was set up. A beautiful garland of fragrant flowers was kept ready; and rose water to wash God's feet on arrival was also kept ready.

The man knew that God is always punctual, so when the clock started chiming twelve, he said, "Hmmm. What happened? God would not disappoint me. He can't be late. Human beings can be late, but not God."

He was puzzled, but decided to wait for another half-hour as a courtesy. Still God did not appear. The guests began sneering, "You fool! You said God was coming. We had doubts. Why on earth would God come and eat with you? Let's go!"

The man cried, "No. Please wait for a few minutes." He stepped into the dining room to check everything. To his dismay a huge black dog was right on the dining table, ravenously eating everything in sight.

"Oh no!" cried the man. "God knew that this dog had come to defile the food. That's why He didn't want to eat it." He grabbed a big club and started beating the dog furiously. The dog cried and ran away.

"Well, what can I do now?" the man asked his guests. "Now neither God nor you can eat the food because it has been polluted by a dog. I know that is why God didn't come."

Slowly everyone left. The host felt so bad that he went into his shrine room and started praying.

After some time he heard a little moaning sound. He opened his eyes and saw God there before him. God was all covered with bandages and slings. Bruises and abrasions blotched His face.

"Oh Lord! What happened?" asked the man. "You must have gotten in a terrible accident."

"It was no accident," said God, "it was you."

"Why do You blame me?"

"I came punctually at noon and started eating all the delicious food. Then you came and beat me. You clubbed me and broke my bones."

"But you did not come!"

"Are you sure nobody was eating your food?"

"Well, yes, there was a horrible big black dog. He was right on the table devouring everything. I beat him and chased him away. I was sure You didn't come because You knew about the dog."

"Who was that dog then, if not Me? I wanted to really enjoy your food. Who else is better to appreciate food and eat plenty of it but a dog? So I decided to come in the form of a dog."

The man fell on the floor and begged God for forgiveness.

Conceptions of God

Our concept of God evolves as our consciousness expands. Many people conceive God as one who is outside of the universe, who lives in heaven, and who punishes the bad and rewards the good. When our consciousness expands, our concept of God and of ourselves grows. When a worshipper identifies himself with the body, he looks upon God as his Master and himself as the servant. When he finds himself as an individual soul, he regards God as the Soul of all souls. When he perceives himself as the Spirit (*Atman*), he realizes that he and God are One. "God," says Plotinus,[4] "is not external to anyone, but is present with all things, though they are ignorant that He is so." (*Enneads*, vi. 9) This experience, only a few can gain, and of those who gain it, none can really describe this state of absolute union with God. The worshipper rises above all limitations, all that is non-self (body, mind and senses), and realizes his oneness with the Divine. In this non-dualistic stage, the devotee realizes that the God he has been worshipping first outside himself, and then as the immanent Being, is really the transcendent Reality.

The seers of the *Upanishads*, in the depth of meditation, realized the supreme immutable being called *Brahman*. The *Upanishads* speak of *Brahman* as the One Eternal Reality. The central vision of the *Upanishads* is *Brahman* (Absolute). Although *Brahman* is beyond thoughts and words, each one of us as *Atman* or as our Self can feel him.

Katha Upanishad describes Self:

"The knowing Self is never born; nor does it die at any time.
It sprang from nothing and nothing sprang from It.

It is unborn, eternal, abiding and primeval.
It is not slain when the body is slain." (1. 2. 18)

There is a similar description in the *Bhagavad Gita:*

"It is not born nor does it die; after having been,
It again ceases not to be. Unborn, eternal, changeless and ancient,
It is not killed when the body is killed." (2. 20)

The Self constitutes the inner reality of each individual. It is without a cause and is changeless. *Katha Upanishad* states further:

"Smaller than the small, greater than the great, the Self dwells in the
Heart of every being. When a person is free from desire, and
His mind and senses are tranquil, he beholds the glory of the Self
And is free from sorrow." (1. 2. 20)

Brahadaranyaka Upanishad declares:

"This Immutable is never seen, but is the seer: it is unheard, but is the hearer;
It is unknown, but is the Knower. There is no other seer but this,
No other hearer but this, no other thinker but this,
No other knower but this." (3. 8. 11)

The Vedanta describes *Brahman*, the ultimate Reality as *Satchidananda,* which means Existence-Knowledge-Bliss Absolute. The entire universe is its manifestation.

The Sufi mystics speak of their Divine Realization in terms of love and unity: "Whoever enters in Love's city finds room only for One and only in Oneness union… In God there is no duality. In that presence, 'I' and 'me' do not exist. 'I' and 'me' and 'you' become one. Since in unity there is no distinction, the quest, the way and the seeker become one."

In the words of Sri Ramakrishna, *Brahman* is beyond knowledge, it is beyond ignorance, and it is beyond *Maya* (Cosmic

Illusion), beyond duality. *Brahman* is not only beyond the drama of worldly life, but also beyond the subtlest philosophical knowledge. It is beyond right and wrong, good and evil.

What *Brahman* is cannot be described. All the revealed scriptures and profound philosophies have been defiled by the limited mind of man. Only one thing has not been defiled, and that is *Brahman*. No one has ever been able to say what *Brahman* is.

Brahman alone exists as the pure Consciousness within and beyond all finite forms of consciousness, and as *Shakti,* the primordial Energy. *Brahman* and *Shakti* are identical, like the fire and its power to burn. (Source: *The Gospel of Sri Ramakrishna*).

The Conceptions of God in ancient India

In the scriptures of India, we come across various conceptions of Godhead: as One without form or One with form and attributes. The Vedic seers worshipped Indra (the rain god), Mitra (the god who regulates the course of the sun), Varuna (the god of the sky), Agni (the god of fire), and Savitr (the solar deity). Even at that stage some of the seers recognized the presence of an indwelling Spirit behind all natural phenomena. A well-known hymn in *Rig Veda* declares: *"Ekam sat; vipra bahudha Vadanti."* This means "The Truth is one, the sages call it by many names." (1. 164. 46)

After the period of the *Rig Veda* the worship of Shiva, Vishnu, Devi, Rama and Krishna and other Hindu Gods became popular. However, the personal aspects of God were recognized as the manifestations of formless Supreme. According to the *Advaita Vedanta*, whatever be the symbol or the personal God one may worship to begin with, the highest goal of spiritual life lies in the ultimate experience of the impersonal, In that highest experience all distinctions between man and God, and between God and the world vanish; the Absolute Consciousness alone remains. It is this impersonal Divine Principle that manifests itself through finite forms and yet remains infinite and formless. *Brhadaranyaka Upanishad* declares:

"It exists in and through the earth, the air, the sun, the moon and the stars;
It controls everything and every being from within; It is the Internal Ruler;
The Immortal Self of the worshipper." (3. 7. 7)

The Relationship between the Self and God

The Self within us is our true nature. It is eternal, unborn and changeless. The *Gita* says, "This, the Indweller in the bodies of all, is ever indestructible." (2. 30) One who experiences the Self becomes immortal and free from fear and misery. What is the relationship between the Self and God? To the mystical theist, the Indwelling Spirit, the Self within is God. To the monist, God is his true Self, as distinct from the individualized, false self, which he takes to be real before the dawn of the highest spiritual knowledge. In trying to realize his real nature, he finds that what he has been calling his own self or personality is only a shadow of reality. The perfection of this ideal is realized when he becomes one with It. Speaking on this point, Sri Ramakrishna observes: "Know yourself and you shall then know God. What is my ego? Is it my hand or foot or flesh or blood or any other part of my body? Reflect well, and you will know that there is no such thing as 'I'. The more you peel an onion, the more you find it to be all skin; you cannot get any kernel at all. So when you analyze the ego, it vanishes into nothingness. What is ultimately left behind is the *Atman* (Self) – the pure *Chit* (Absolute Consciousness). God appears when the ego dies."

Worship of God as the Mother

God can be worshipped in many ways: as the Master, as the Mother, as the Divine Child, as the Father, or as the Divine Lover. (For details, *see* chapter 16. Attitudes (Bhavas) of Devotees.) Just as a baby is nursed and nourished by its mother, God creates, supports, and nourishes all living beings. To look upon God as the Mother is the most natural way. It is also the most sublime attitude. There is greater freedom and spontaneity in a devotee's relationship with God as the Mother. As Sri Ramakrishna used to say, just as a child can force its demands on its mother, so also can the devotee force his demand on God.

The concept of God as the Mother prevailed in many countries in ancient times. Hindus in India worship God as Kali, Durga, Lakshmi, Sarasvati, Devi and so on. In Egypt She was known as Isis, in Babylonia and Assyria as Ishtar, in Greece as Demeter, and in Phrygia as Cybele. The Romans also worshipped Cybele as the mother of gods. Judaism and Islam put an end to Mother worship in the Middle East. However, Catholics venerate Virgin Mary as the Mother of God.

In the West a woman is looked upon as a sweetheart or wife. The mother does not command much respect and love as in a Hindu home. If men in the West worshipped God as the Mother, they would be less sex-conscious and be more spiritual. There would be stronger family bonds and greater domestic peace.

In India the worship of God as the Mother has an unbroken tradition right from the Vedic times. In later centuries, a whole body of literature called the *Tantras* was dedicated to the philosophy and worship of the Divine Mother. With the advent of Sri Ramakrishna, Mother worship has undergone a rejuvenation. He looked upon Mother Kali as the creative Power of the universe. To him She is inseparable from *Brahman*; the immutable Reality is *Brahman*; when It manifests as the universe It is Kali. Sri Ramakrishna thus elevates the worship of the image of Kali to the adoration of the Infinite.

The Mother power or the Divine Energy has many forms and symbols. She is symbolized variously as Durga (the Goddess of Destruction of all evil forces and tendencies), as Lakshmi (the Goddess of Wealth), and as Sarasvati (the Goddess of Knowledge). In the form of Kali, She is represented as the power of creation, protection and destruction, and as the power in which all things rest after dissolution. She stands on the still and prostrate form of Shiva, the representation of the Absolute. This is symbolical of the entire cosmic process with transcendental Reality as its basis.

Reality is beyond both life and death, and as such, the devotee should neither cling to life nor be afraid of death. He should rise above both pleasure and pain and other dualities, to the transcendental plane from where he can say, in the words of Swami Vivekananda's *Hymn to the Divine Mother*: "The shade of death and immortality – both these, O Mother, are Thy grace supreme!"

Avatar (Incarnation)

Avatar is a Divine Incarnation. God incarnates Himself in different ages under different forms and names such as Rama, Krishna, Buddha, Christ, Muhammad and Ramakrishna. In the *Bhagavad Gita*, Lord Krishna says:

> "Whenever there is a decline of righteousness, O Arjuna,
> And rise of unrighteousness, then I manifest Myself,
> For the protection of the good, for the destruction of the wicked,
> And for the establishment of righteousness, I am born in every age."
>
> (4.7,8)

Swami Vivekananda says, "Blessed are those races which have a 'God-man' to worship. The Christians have such a 'God-man' in Christ; therefore, let them cling close to Christ . . . That is the natural way to see God – to see God in man . . . God cannot be worshipped; He is the immanent Being of the universe. It is only His manifestation as man that we can pray to."[5]

According to the *Gospel of Sri Ramakrishna*:

"One can taste devotion and love of God only through His Incarnations. In an Incarnation of God one sees, at the same, the sun of knowledge and the moon of love. God no doubt dwells in all, but He manifests Himself more through human beings than through other beings. Man can think of God, he can think of the Infinite, while other beings cannot. Fire exists in all beings, in all things, but its presence is felt more in wood. So also in the Incarnation there is a greater manifestation of God than in other persons. He exhorts people to self-surrender to God. Following the path of devotion, one realizes everything through His Grace – both knowledge and Supreme Wisdom.

"Ordinary people do not recognize the advent of an Incarnation of God. Only a few of His intimate disciples can recognize Him. That Rama was both *Brahman* Absolute and a perfect Incarnation of God in human form was known only to twelve *rishis* (seers).

"It is the *Sakti*, the Primal Energy that is born as an Incarnation. Though an Incarnation has a human body, He is always united with God. Though established in *Samadhi*, He can again descend to the worldly planes."

The worship of an *Avatar* is a special category of *Bhakti*. A devotee accepts any of the Incarnations as his spiritual Ideal or personal God through whom he tries to establish contact with the Absolute. He sees Divinity and Divine attributes in the Incarnation. The human form is only a symbol of the Divine.

Rama is one of the incarnations of Vishnu. He is the embodiment of truth and duty. The ordinary worshipper stresses His lovely form and noble attributes. But the illuminated devotee sees Him immanent in all and prays:

"Thou art the embodiment of the highest virtues. Thou art the indweller, the Supreme Being. Thou art the greatest refuge and saviour of mankind. (*Valmiki Ramayana*[6], 6. 117. 14. 17)

"Thou art the stainless, changeless, indestructible, pure and eternal wisdom and truth." (*Adhyatma Ramayana*[7], 1. 5. 56)

Krishna is a complete *Avatar*. He is worshipped in His various forms. Devotees see in Him the highest ideal of Divine Love. Realizing the vision of Sri Krishna's Universal Form, Arjuna makes obeisance to Him, saying:

"Salutations to Thee on every side, O Lord, Thou art everything;
Infinite in power, infinite in prowess, Thou pervadest all;
Thou art all." (*Bhagavad Gita*, 11. 40)

Ways of using an Incarnation.

There are three ways of using an Incarnation in spiritual practice. The first is to use the Incarnation's life as a model for one's own existence. Rama's devotion to truth and duty, Krishna's ideal of Divine love, and the compassion of Christ are exemplary guideposts for the aspirants.

The second is to identify, study and contemplate particular incidents or aspects of the Deity's life and use them as models.

The third is to read and study the teachings of the Incarnate and follow them in one's life. The teachings of Lord Krishna in the *Bhagavad Gita*, the *Gospel of Christ*, and the teachings of other Incarnations are lighthouses in the spiritual world.

Avatar's Mission

An *Avatar* brings the light and power of the Supreme Truth to a vast mass of humanity. He embodies not only the inner perfection of the Spirit, but also exercises all the powers and glory of the Divine.

An *Avatar* descends on the earth with the special mission of liberating mankind from ignorance and bondage. *Avatar*s like Krishna, Buddha, Christ, Mohammed, and Zoroaster are saviours of humanity and redeemers of all fallen souls.

The *Avatar*s have a message for the whole world. They preach universal love as the highest attainment, which alone can establish unity, harmony and brotherhood in the world.

God has exhibited omnipotent powers through the *Avatars*. Apart from performing miracles, they release a world-redeeming spiritual force by which they transform people steeped in ignorance and sin into veritable angels of purity, wisdom, love and peace. Their light and power illuminate whoever comes in contact with them even in thought and meditation.

Sri Ramakrishna often speaks of Divine Reality as a boundless golden meadow, blocked from view by the apparent wall of space and time, substantiality and separation. The God-man or *Avatar* is like a vast opening in that wall, through which millions of sincere human beings can pass without the need of the superhuman efforts at contemplation exerted by the saints.

An *Avatar* is a Divine human being who is nothing other than God's light. For the *Avatar*, there are no fundamental considerations of body and soul, nor is there any special path. For him there is no *Karma*, no limiting or binding impressions from the history of personal and collective embodiments. The human body and mind of the *Avatar* is an opaque covering; beneath this veil there is no individual soul, no eternal facets of the Divine, but instead there resides the complete Divine Reality with infinite facets.

No scripture will ever be adequate to the ineffable teachings that flow through the *Avatar*. The primordial power of the Pure Consciousness plays freely through the teachings, the daily life and even the slightest movements of the *Avatar*. It is easy and delightful to meditate on an *Avatar*. Burning sunlight becomes gentle and healing when it manifests as moonlight. An *Avatar* is simply God at play within God. (Source: Lex Hixon, *Great Swan*)

Unity behind Diversity

Devotees with a strong sectarian bias very often hold that salvation lies only through the worship of their particular Deity or Incarnation or through devotion to their formless God following the teachings of only certain prophets. Along with such sectarian devotees there are liberal souls who, while yielding to none in their love and faith for their chosen ideal, look upon all Divine Personalities as the same Truth.

This unity in the midst of diversity has been a basic feature of Hinduism all along its unbroken course. *Manu Smriti*, an ancient Sanskrit Scripture of India proclaims:

> "One ought to know the Supreme Spirit who is the Ruler of all,
> Subtler than the subtlest, of resplendent glory,
> And capable of being realized only by meditation.
> Some call Him Agni (Adorable), others call Him Manu (Thinker);
> And others Prajapathi (Lord of Creatures);

> Some again call Him Indra (the Glorious),
> Others Prana (the Source of Life), and still others
> The Eternal *Brahman* (the Great)." (12. 122 - 23)

Upon recognizing the universal aspect of a Divine Being, the followers of all religions and creeds may join their hands and hearts in common worship. This may serve as a great bond for uniting the truly religious- and spiritually-minded souls in all lands, and for making them work together in a spirit of brotherhood and fellowship, service and cooperation for the common weal.

> "May He – the One-without-a-second – who, though formless
> And without any purpose of His own, produces various forms
> By means of His manifold powers, from whom the universe come into
> Being in the beginning of creation, and to whom it returns in the end,
> Endow us with good thoughts." (*Svetasvatara Upanishad*, 4. 1)

God is One: The Unity of Religions – Sri Ramakrishna's Demonstration

In Sri Ramakrishna's life (1836-1886), we find an actual proof for the unity of all religions. When he became the priest of Kali Temple at Dakshineswar, a village about four miles north of Calcutta, he poured his soul into Kali-worship. The worship in the temple intensified Sri Ramakrishna's yearning for a living vision of Kali, the Mother of the Universe. He began to practice intense meditation during nights in the deep jungle near the temple. Most of the time he was unconscious of the world. He almost gave up food and sleep left him altogether. But he did not have to wait very long. He describes his first vision of the Mother:

> "I was overpowered with a great restlessness... I could not bear the separation from Her any longer... Suddenly my glance fell on the sword that was kept in the Mother's temple. I determined to put an end to my life. When I jumped up like a madman and seized it, suddenly the blessed Mother revealed Herself. The buildings with their different parts, the

temple and everything else vanished from my sight, leaving no trace whatsoever and in their stead, I saw a limitless, infinite, effulgent Ocean of Consciousness."

The first glimpse of Divine Mother made him more eager for Her uninterrupted vision. He wanted to see her both in meditation and with eyes open. With greater intensity he continued his spiritual practice. His visions became deeper and more intimate. Even while retaining consciousness of the outer world, he would see the Divine Mother as tangibly as the temples, the trees, the river and the men around him.

Later Sri Ramakrishna began to worship God by assuming the attitude of a servant toward his master. He imitated the mood of Hanuman, the monkey devotee of Rama. He lived like a monkey and meditated on Hanuman. After a short while he was blessed with a vision of Sita, the Divine consort of Rama, who entered his body and disappeared.

Sri Ramakrishna was passing through the most exalted rapture of divine love. At this time he came to Dakshineswar, and met an elderly *Brahmin* nun, adept in the *Tantrik* and *Vaishnava* methods of worship. Very soon a tender relationship sprang up between Sri Ramakrishna and the *Brahmani*, where she looked upon him as the baby form of Lord Krishna and he upon her as his mother. After watching his ecstasy and his mad yearning day after day, she came to the conclusion that only an incarnation of God was capable of such spiritual manifestations. She proclaimed openly that Sri Ramakrishna, like Sri Chaitanya, was an incarnation of God.

At the bidding of the Divine Mother Herself, Sri Ramakrishna accepted the *Brahmani* as his Guru and practiced the *Tantra* and *Vaishnava* disciplines. Within three days he beheld everywhere the power and beauty of the Divine Mother; the whole world appeared to him as pervaded with the bliss of God-consciousness. He saw in a vision the Ultimate Cause of the universe as a huge luminous triangle. He heard the *Anahata* sound, the great sound *Om*. He had a vision of the divine *Maya*, the inscrutable Power of God. Above all, he experienced the awakening of the *Kundalini Sakti*.

After completing the *Tantrik* practice, Sri Ramakrishna practiced the disciplines of *Vaishnavism*, a Hindu religion of *Bhakti*. He was blessed with a vision of Ramachandra and realized that Rama pervades the whole universe as Spirit and Consciousness, as its Creator, Sustainer and Destroyer, and as the transcendental *Brahman*.

He then wanted to attain union with Sri Krishna. He regarded himself as one of the *Gopis* (milkmaids) of Brindavan. But Krishna began to play with him the old tricks he had played with the *Gopis*. He would reveal Himself occasionally, always keeping a distance. As advised in the *Vaishnava* scriptures he turned his prayer to Radha and sought her grace to realize Sri Krishna. Within a very short time he enjoyed her blessed vision. He saw and felt the figure of Radha disappearing into his own body. Now one with Radha, he manifested the great ecstatic love. The enchanting form of Krishna appeared to him and merged in his person.

Totapuri, a wandering monk arrived at Dakshineswar toward the end of 1864. He was an *Advaita* (Non-dualistic) *Vedantist*. He initiated Sri Ramakrishna into the *Vedantic* discipline. To the utter astonishment of his Guru, Sri Ramakrishna attained *Nirvikalpa Samadhi* in a single day and realized his identity with *Brahman*. Sri Ramakrishna discovered: *Maya* is none other than Kali, the Divine Mother; She is the primordial Divine Energy, Sakti; She is the Mother of the Universe, identical with the *Brahman* of *Vedanta* and with the *Atman* of Yoga. The Divine Mother asked him not to be lost in the featureless Absolute but to remain in *Bhavamuka*, on the threshold of relative consciousness, the border line between the Absolute and the Relative.

The Knowledge of *Brahman* in *Nirvikalpa Samadhi* had convinced Sri Ramakrishna that the Gods of different religions are but so many readings of the Absolute. He understood that all religions lead their devotees to one and the same goal. He now wanted to explore other religions.

Towards the end of 1866 he began to practice Islam under the direction of an Islamic Guru. After three days he saw the vision of a radiant figure. This figure gently approached him and finally merged in Sri Ramakrishna. Thence he passed into communion with *Brahman*.

Eight years later, Sri Ramakrishna had an irresistible desire to learn the truth of the Christian religion. He began to listen to readings from the *Bible*, by one of his devotees. The life and teachings of Jesus fascinated him. One day Sri Ramakrishna was sitting in the parlor of a devotee's house, when his eyes became fixed on a painting of the *Madonna and Child*. Intently watching it, he became overwhelmed with divine emotion. The figures in the picture took on life and the rays of light emanating from them entered his soul. Christ possessed his soul. He cried out, "O Mother! What are You doing to me?" For three days he did not set his foot in the Kali temple. On the fourth day, as he was walking in the temple garden, he saw coming towards him a person with beautiful large eyes, serene countenance and fair skin. Sri Ramakrishna cried out: "Behold the Christ, who shed His heart's blood for the redemption of the world, who suffered a sea of anguish for love of men. It is He, the Master Yogi, who is in eternal union with God. It is Jesus Love Incarnate." Jesus merged in Sri Ramakrishna, who then went into *Samadhi* and communed with *Brahman*. Thus he experienced the truth that Christianity, too, was a path leading to God-Consciousness.

Sri Ramakrishna accepted the Divinity of Buddha and used to point out the similarity of Buddha's teachings to those of the *Upanishads*. He also showed great respect for the *Trithankaras* who founded Jainism and for the ten Gurus of Sikhism. Thus he became a Master who could speak with authority regarding the ideals of various religions of the world. "I have practiced," said he, "all religions – Hinduism, Islam, Christianity – and I have also followed the paths of the *Saktas* (worshippers of Sakti), *Vaishnavas* and *Vedantists*. I have found that it is the same God toward whom all are traveling, though along different paths." It is thus Sri Ramakrishna actually proved by direct experience the unity of different religions. May the people of the world, by the Grace of God, realize this eternal truth and live as one Divine family. (Source: *The Gospel of Sri Ramakrishna,* 19)

REFERENCES

1. One of the Incarnations of God; founder of Buddhism in India, which later spread to other countries.

2. *Brahmo Samaj* is a theistic organization founded by Raja Rammohan Roy of India.

3. Swami Nikhilananda, tr., *The Gospel of Sri Ramakrishna*, Madras: Sri Ramakrishna Math, 1986, pp.149-50.

4. Plotinus, 205-270 was the founder of Neoplatonism that refers to the influence of Plato in western philosophical, religious and political thinking. The essays that he wrote for circulation among his pupils were collected in six sets of nine known collectively as the *Enneads*.

5. *Religion of Love*, Calcutta: Udbodhan Office, 1970.

6. One of the great Epics of India, composed by the sage Valmiki. It narrates the story and glory of Sri Rama.

7. A scripture dealing with the life of Sri Rama, an Incarnate, and harmonizing the ideals of *Jnana* (Wisdom) and *Bhakti* (Devotion).

4

Bhakti Yoga

Love

As oil is present in every part of the olive, so love permeates every part of creation. To define love is very difficult. Words cannot fully describe the experience of love, just as words cannot fully describe the flavour of an orange. We have to taste the fruit to know its flavour.

Love is a universal emotion; its expressions are distinguished by the nature of feelings through which it is expressed. A father may express motherly love, a mother may express friendly love, and a lover may express Divine love. Every reflection of love comes from one Cosmic Love.

Cosmic Love or Divine Love is the soul's love for the Supreme Spirit. Divine Love is joy. The more you meditate and seek God with a burning desire, the more you will feel that Love in your heart. Then you will know that love is joy, and joy comes from God's bliss, the essential and ultimate nature of the Spirit. *Taittiriya Upanishad* says, "From joy we have come, in joy we live and have our being, and in that sacred joy we will one day melt again." (3. 6. 1)

Narada's Definition of *Bhakti*

Narada, the Divine Sage, defines *Bhakti* as intense Love for God (*N.B.S.*, 2). Intensity is one of the characteristics of Divine Love. The soul's intense thirst produces Divine response. The intense love that the sage speaks of refers to the love that arises in the heart of a devotee when he has become intoxicated with Divine Love. Such love is the same as experiencing God-consciousness. Intense love is more than having a vision of God in ecstasy. At a higher stage, love, lover and the beloved become one. Then God is experienced as transcendental Divinity. Intense love is *Prema*. Sri Ramakrishna defines *Prema* as: "the intense Love for God that makes one forget the world and forget even one's own body, that is, rise above physical consciousness." In this experience of inexpressible bliss the ego is completely lost.

Here are some illustrations from the saint and seers who attained complete union with the Lord through such intense devotion to God. Prahlada, the great devotee of Sri Krishna, when absorbed completely in the consciousness of God, found neither the universe nor its cause; all was to him one Infinite, undifferentiated by name and form.

So it was with the *Gopis*, the milkmaids of Brindavan. As soon as they lost themselves in their absorbing love for Krishna, they realized their union with Him and became Krishnas.

Sri Chaitanya, the sixteenth century saint of Bengal was a great lover of God. He used to experience three moods. In the inmost mood he would be absorbed in *Samadhi*, unconscious of the external world. In the semiconscious mood he would dance in ecstasy but could not talk. In the conscious mood he would sing the glories of the Lord. In the life of Sri Ramakrishna we find that he would become absorbed in God many times during the day. Then he would experience unitary consciousness; and later, as he would come back to normal consciousness, he would speak of God as the blissful Mother.

Other Definitions

Many of the saints and scriptures from the early days hold the opinion that *Bhakti* is firm love:

- Prahlada describes *Bhakti* as: "That abiding love which the non-discriminates have for the fleeting objects of the senses, may I have the same sort of love for Thee, may not that joy vanish from my heart as I think of Thee." (*Vishnu Purana*, 1. 20. 19)

- Uddhava, a great devotee of Sri Krishna, declares that *Bhakti* is a loving sense of possession. (*Naradapancaratna*)
- The constant flow of mind, brimming with love towards the Lord, without any selfish desire is *Bhakti*. (*Ibid*)
- Loving contemplation of the Lord without break or motive of gain – free from the overpowering influence of *Jnana* and *Karma* – is the highest *Bhakti*. (*Bhaktirasamrtasindhu*)
- *Bhakti* is the adoring service of the Lord, centring the mind on Him, and expecting no other gain here or hereafter. (*Gopalapurvatapini Upanishad*, 2. 1)
- *Bhakti* is a state of mind in which, the mind, being melted by the force of spiritual discipline, constantly flows towards the Lord. (*Bhaktirasayana*, 1. 3)
- Madhavacharya describes *Bhakti* as: "Firm love for God more than for anything else, with full consciousness of His glory and magnitude." (*Mahabharatam: Tatparyanirnaya*, 1. 86, 107)

The Christian saints and mystics reflect the same ideas about love for God:

- Saint Paul says, "The best is to be one with God."
- Ruysbroeck, the Dutch mystic, says, "We have lost ourselves, and been melted away into the Unknown Darkness."
- According to the Neo-Platonists, the highest stage of union with God cannot be realized by thought, and is possible only in a state of ecstasy in which the soul transcends its own thought, loses itself in the soul of God, becomes one with God.
- Dionysius, a saint of the second century, says, "It is the nature of love to change a man into that which he love."
- Averroes, the Moorish philosopher of Spain, says, "The individual soul becomes one with the Universal Spirit or is absorbed in It."

- The German mystic Eckhart says, "The soul in her hot pursuit of God becomes absorbed in Him."

The sayings of some Sufi mystics reflect the same ideas about Divine Devotion:

- Jalaludin Rumi says, "Then shall we rise from the angels and merge in the Nameless."
- Another Sufi says:
- When a man becomes annihilated from his attributes, he attains perfect subsistence. He is neither far nor near... neither separated nor united; He has no name or sign, or brand or mark." (Amir Ali, *Spirit of Islam,* pp. 172, 213)
- The Sufi martyr Al Hallaj says, "I am the Truth, I am He whom I love, and He whom I love is I."
- Nazir, a seventeenth century Sufi, says:
  "Then from duality I passed into Oneness
  And all illusions vanished like a phantom show.
  Since then, whatever I see and wherever I cast my glance,
  I see Him alone, none other. The Muslim and the Hindu
  And the Jew have all become the same to me –
  Differences have vanished in the experience of the One."

The Definitions of a few ancient Sages quoted by Narada in his *Narada Bhakti Sutras* are as follows:

- Vyasa, the compiler of the *Vedas* (the revealed scriptures) and the *Puranas* (auxiliary scriptures), defines *Bhakti* as devotion to acts of worship and the like. (*N.B.S.*, 16)

In this definition Vyasa lays stress on worship and such acts that keep our minds fixed on God. Worship includes rituals with offering of flowers, fruit, water, light, incense, and so forth. It also includes mental worship, as well as *Japa* (chanting the Name of the Lord).

The sage Garga defines *Bhakti* as devotion to hearing and praising the name of God. (*N.B.S.*, 17) Garga emphasizes the importance of speech as part of the practice of devotion. A devotee must delight in listening to divine teachings of sages and saints, and the stories of Incarnations. He must also delight in studying devotional scriptures, and in singing God's glories.

Sri Chaitanya (1485–1533), the greatest medieval saint of Bengal (India), in one of his hymns to the Lord, says:

> "Chant the name of the Lord and His glory unceasingly
> That the mirror of the heart may be wiped clean
> And quenched the mighty forest fire of worldly lust
> Raging furiously within…"

Ramprasad, an eighteenth century saint of Bengal, realized his union with God entirely by singing songs, which he had composed in praise of the Divine Mother. In the *Bhagavatam*[1] we read:

> "Wonderful is the teacher, Sri Krishna, wonderful are His deeds.
> Even the utterance of His holy Name sanctifies him
> Who speaks and who hears."

The sage Sandilya defines *Bhakti* as devotion without prejudice to delight in the Atman. (*N.B.S.*, 18) Since God is the experience of one's innermost Self, devotion to God makes a devotee experience delight in the Self. This experience is the surest indication of his progress. Devotion to God and devotion to the inner most Self is the same, because the Self in each one of us is a part of God – the Cosmic Self.

Harmonizing Various Definitions

After quoting the definitions of *Bhakti* given by the sages Vyasa, Garga and Sandilya, Narada sums up, harmonizing all those definitions:

> "When all thoughts, all words and all deeds are given up to the Lord,
> And when the least forgetfulness of God makes one intensely miserable,
> Then love has begun." (*N.B.S., 19*)

In this *Sutra*, Narada points out the very essence of *Bhakti*. Narada does not limit *Bhakti* to rituals, chanting of sacred names, and similar devotional practices only. Complete self-surrender is the prime characteristic of *Bhakti*. Self-surrender is an all-inclusive spiritual discipline. In commenting on Patanjali's *Sutra* on *Isvara Pranidhana* (total surrender to God) (*Yoga Sutras,* 2. 45), Bhoja says, "*Pranidhana* is that sort of *Bhakti* in which, without seeking results such as self-enjoyment and so forth, all actions are dedicated to the Lord, who is the Teacher of teachers."

Ramanuja, a great Eleventh Century saint of Southern India, says in his commentary on the first *Sutra* of the *Vedanta Sutras* (the ancient text of Vedanta philosophy) that the Lord is the very source of joy to the devotee, and devotion to him is a "constant remembrance, flowing like an unbroken stream of oil poured from one vessel to another." To a true devotee even a few moments of forgetfulness of the Lord are moments of despair. He can only regain his peace and balance when he remembers the Lord through worship, singing His glories, or contemplation upon the Self. Thus Narada's definition of *Bhakti* is an integral declaration that encompasses all other definitions made by the other sages.

Devotion

Each individual possesses the capacity for devotion, however this needs to be refined and directed to proper objects. When directed to ignoble objects, it is called attachment. When it is diverted from these objects and fixed on God, it is called devotion. The quality of devotion to God is not strained. It is effortless and natural. God does not command us to love Him. We love Him because it is our inherent nature.

Devotion to God should be pure – free from bargaining. Sarmad, a seventeenth century Sufi says:

"From God everyone seeks wealth and salvation,
Or craves the boon of a moonfaced beauty,
My heart leans toward neither this nor that,

It seeks only union with the Beloved."

Grades of Devotion

Devotion differs in its modes according to the mental level of the devotees. Sri Ramana Maharishi, the twentieth century sage of Arunachala of Southern India, and the ancient lore mention two distinct grades of devotion. They claim that crude minds cannot theoretically grasp the teaching that ultimately God is to be realized as the true Self. These minds are devoted to God with a sense of being different from and subject to Him. They see God as being anthropomorphic, thinking of Him as a superior kind of man, and they look forward to personal rewards.

The ripe devotee, on the other hand, considers himself as nothing, and God as everything, and ultimately wins an egoless state through God's Grace. The goal is attained by the devotee through self-surrender. (For details *see* chapter 25. Cultivation of Bhakti: Positive Practices IV.)

Bhakti Yoga

Bhakti Yoga is the Yoga of Devotion to God. It leads to the same goal as all the other branches of Yoga. It is especially suited for those who are emotional in their nature and have the feeling of love.

The word "*Bhakti*" means devotion, while Yoga signifies union of the individual soul with God. Hence *Bhakti Yoga* is the method of devotion by which true communion of the soul with the Supreme Consciousness is accomplished. It shows how even the ordinary feelings of a human heart, when directed toward God, can become the means of attaining spiritual oneness with the Cosmic Spirit.

Many people are satisfied with performing rituals, formal worship, and observing various ceremonies at stated times of the day, week, month, or on religious festival days. Such observance lasts for a few moments at a time; and at all other times awareness is completely immersed in and carried away by worldly thoughts. With *Bhakti Yoga* the sacred emotion is made lasting, persisting and enduring. The Yogi is yearning to be filled with holy and pure emotion towards

God. Yoga means to yoke onto, or to put two together. It puts the Yogi in the company of God. An ordinary devotee has a fleeting devotion to God in certain stated moments of his life, while the *Bhakti* Yogi has his whole time, attention and eagerness yoked with God. Every soul has an intrinsic urge to love God, and *Bhakti* Yoga enables the soul to discover that eternal Divine Love and to dissolve into It. However, several impurities or obstacles in the mind obscure this ray of Divine Love. These obstacles include egoism, selfishness, attachments, aversions, impressions of past thoughts and experiences, effects of past actions, and so on. (For details, see Chapter *18. Obstacles and Remedies,* below.) *Bhakti* Yoga is a means for removing these obstacles and purifying the mind. The practice of *Bhakti Yoga* cultivates your emotions and expands your heart, enabling you to regard every being as your Self, making you less attached and selfish, and more saintly and divine.

Worldly relationships and sense-objects cannot quench your inherent thirst to love, because you are seeking an eternity that transcends the relative world of time and space. *Bhakti Yoga* aims at training your sentiments, pulling you out of worldliness, and directing your mind to the transcendental Divinity. As a result, you experience mystic joy flowing from Divine Love.

Bhakti in Religious Literature

While pre-Aryan Indian civilizations were pro-worship and practiced the *Bhakti* cult, the *Vedas* do not mention the word *Bhakti* (in the sense of concentrated devotion to God). The Vedas focus on worship by means of *Yagna* – offering to the fire, where various *Devas* (or Gods) such as Indra and Mitra are worshiped. Likewise, the *Upanishads* focus on *Jnana Yoga*, investigating the *Atman* or Self, which is identical with *Brahman* – non-dual Reality.

It is in the *Bhagavad Gita* that *Bhakti Yoga* is found fully developed. The concept of *Avatar* is found only in the *Gita* and in later literature. With the decline of *Vedic* rituals and the rise of temple-worship, *Bhakti* took a definite hold on the minds of the masses. The advent of the *Puranas* made worship of Shiva, Vishnu, Sakti and *Avatar*s very popular. The growth of the *Tantra* literature also contributed to the growth of *Bhakti*.

Furthermore, the philosophies of Vedanta, Samkhya, and Yoga contributed to a higher development of the ritualistic form of *Bhakti*. Spiritual growth evolved from the idea of propitiation of God to the idea of God-realization through *Bhakti Yoga*.

Sri Sankaracharya firmly observes that *Para Bhakti* (Supreme Devotion) and Supreme *Jnana Yoga* are but one. While the initial procedures vary, the ultimate outcome is the same. In *Bhakti*, the devotee thinks that God is different from him and God is to be realized by intense devotion and total self-surrender to Him. In *Jnana Yoga*, the investigation is about one's own Self. The aspirant attains Self-realization through study and contemplation of the teachings of *Vedanta*, as well as understanding and meditating on the Self. However, common requirements of both paths include extreme eagerness, supreme purity, intense effort and one-pointed attention to achieve the ultimate goal.

Despite *Bhakti Yoga* being of later origin than *Jnana Yoga*, it is a living force in the spiritual life of India. Since the post-Buddhist era, India has produced many great *Bhakta* Yogis like Ramanujacharya, Madhvacharya, Chaitanya, Tukaram, *Alvars* (Vaishnava Tamil saints), *Nayanmars* (Tamil Saiva saints), and a host of others. Other religions like Christianity and Islam have also produced many great mystic saints.

The greatest *Bhakta* Yogi of modern times is undoubtedly Sri Ramakrishna, the God-man of nineteenth century. He is an outstanding example of one who had attained *Nirvikalpa Samadhi* of *Jnana Yoga*, after reaching the greatest height of *Bhakti*. He is also the only saint who actually demonstrated in his own life the unity of all religions and faiths. His life is a rare illustration for the basic truth that all paths lead to the same goal (For details *see* chapter 3. What is God? and also the next sub-title.).

Bhakti Yoga–**Easiest**

When a devotee's mind is concentrated and one-pointed; consequently meditation becomes easy for him. Continuous remembrance of the Divine Ideal leads to unceasing meditation, and ultimately lifts the soul into *Samadhi*, where it realizes God. He then sees the presence of God everywhere and enjoys unbounded peace and happiness at all times. It is for this reason that *Bhakti Yoga* is considered to be the easiest of all methods. What a *Raja Yogi* attains only after years of practice, a *Bhakta* accomplishes in a short period through intense devotion and love. That which a *Karma Yogi* finds so difficult to achieve, a *Bhakti Yogi* attains easily by surrendering everything including himself to the Lord. In the words of Sri Ramakrishna:

> "*Bhakti Yoga* is the religion for this age. But that does not mean that the lover of God will reach one goal and the *Jnani* and *Karma* Yogi another. It means that if a person seeks the Knowledge of Brahman he can attain It by following the path of *Bhakti*, too. God, who loves His devotee, can give him the Knowledge of *Brahman*, if he so desires.
>
> "But the *Bhakta* wants to realize the Personal God endowed with form and talk to Him. He seldom seeks the Knowledge of *Brahman*. But God who does everything at His pleasure, can make His devotee the heirs to His infinite glories if it pleases Him. He gives His devotee both the Love of God and the Knowledge of *Brahman*.
>
> "By realizing the Divine Mother of the Universe, you will get Knowledge as well as Devotion. You will get both. In *Bhava Samadhi* (Ecstasy) you will see the form of God, and in *Nirvikalpa Samadhi* (Oneness with God) ego, name and form do not exist." (*The Gospel of Sri Ramakrishna*, 1986, p.468)

Path of Love: Only One can Walk

Kabir (Fifteenth Century), a great poet-saint of Northern India, said:

> When I was rapt in myself, O beloved, Your face was hidden from me.
> Now that I see You, I am no longer I.
> I now know this Path of Love is much too narrow,
> Two cannot walk here, only one.

What we learn from great scriptures like the *Bhagavad Gita* is this: you have to know your own Self. You cannot have ego and the higher Self at the same time. You have to lose your ego in the Supreme Self. When you give yourself to God, you become nothing but God, and then God manifests in you. Illusion exists only as long as you allow it to exist. True devotion or sacrifice is giving yourself to that which you believe. When you completely give yourself to God, you receive God completely, and God becomes all that you experience.

REFERENCE

1. *Bhagavatam* is the most important among the *Puranas*, Indian devotional literatures. By means of stories from the lives of *Avatar*s, sages, devotees and kings, it popularizes the truths contained in the *Vedas*.

5

Aims of Bhakti Yoga

God-vision

The primary aim of *Bhakti Yoga* is God-realization. God cannot be realized without purity of heart. A devotee receives the Grace of God by subduing the passions: lust, anger, greed, and egoism, and thus gets God-vision.

People have realized God in various ways. Some attain perfection through much austerity, worship and devotion. Some are born as *Nityasiddhas* – eternally perfect, for example, Narada and Sukadeva. Some attain perfection all of a sudden, like unexpectedly receiving a great fortune; and some realize God in a dream by Divine Grace. Sri Ramakrishna describes the signs or characteristics of the devotee who has received God-vision: he becomes like a child; he is beyond the three *Gunas* (qualities) of *Sattva* (peace), *Rajas* (passion) and *Tamas* (inertia); and he is unattached to any of them. He behaves like a ghoul, for he maintains the same attitude towards things holy and unholy. Again like a mad man, he sometimes laughs and some times weeps. At times he sits motionless like an inert thing. Further he is firmly convinced that he is a machine and God is the operator, and that God alone is the Doer and all others are His instruments. Even the leaves move because of God's will.

A God-realized person becomes like a dry coconut. He is utterly free from body-consciousness. He does not seek comforts of the body. He roams about in the world as a *Jivanmukta*, one liberated while alive. (For details *see* chapter 32. Jivanmukta.)

One of the signs of God-vision is joy; there is no hesitancy in the realized person. Like the ocean, waves and sounds are on the surface; profound depths are below. Sometimes he remains speechless, because he sees God within and without; sometimes like a child, without any attachment, he wanders about unconcerned. He acts like a boy, indulging in frivolity; sometimes like a young man he works and teaches with the strength of a lion. (*The Gospel of Sri Ramakrishna*)

Vision of God: Sri Ramakrishna's Experience

As a priest of Kali Temple at Dakshineswar, Sri Ramakrishna had intense yearning for a living vision of Kali, the Mother of the Universe. How He realized Kali, Rama, Krishna, Brahman, Sakthi, Allah and Jesus Christ has been described in chapter 3. (What is God?). Here he describes his other visions and experiences:

"I used to worship the Deity in the Kali temple. It was suddenly revealed to me that everything is Pure Spirit. The utensils of worship, the altar, the door frames – all Pure Spirit. Men, animals and other living beings – all Pure Spirit. Then like a madman I began to shower flowers in all directions. Whatever I saw I worshipped.

"One day, while worshipping Siva, I was about to offer a *Bel*-leaf on the head of the image, when it was revealed to me that this *Virat*, this Universe, itself is Siva. Another day I had been plucking flowers, when it was shown to me that each plant was a bouquet adoring the Universal Form of God. When I see a man, I see that it is God Himself walking on earth, as it were, rocking to and fro, like a pillow floating on the waves.

"God talked to me. It was not merely His vision. Under the banyan tree I saw Him coming from the Ganges. Then we laughed so much! By way of playing with me He cracked my fingers. Then he talked to me.

"For three days, I wept continuously. And He revealed to me what is in the *Vedas*, the *Puranas*, the *Tantra* and other scriptures.

"One day in the Kali temple Haladhari and Nangta (Totapuri) were reading the *Adyatma Ramayana*[1]. Suddenly I had a vision of a river with woods on both sides. The trees and plants were green. Rama and Lakshmana were walking along, wearing their shorts. One day, in front of the *kuthi* (bungalow), I saw Arjuna's chariot. Sri Krishna was seated in it as the charioteer. Another day, when listening to *kirtan* (devotional songs) at Kamarpukur, I saw Gauranga (Chaitanya) in front of me. At that time a naked person, emerging from my body, used to go about me. I used to joke with him. He looked like a boy and was a *Paramahamsa* (highest monk).

"One day God showed me the *maya* of Mahamaya. A small light inside a room began to grow, and at last it enveloped the whole universe.

"Further, He revealed to me a huge reservoir of water covered with green scum. The wind moved a little of the scum and immediately the water became visible, but in the twinkling of an eye, scum from all sides came dancing in and again covered the water. He revealed to me that the water was like *Satchidananda* (Existence-Knowledge-Bliss Absolute) and the scum like *Maya*.

"God reveals the nature of the devotees to me before they arrive. The Divine Mother also showed me in a vision, the five suppliers of my needs: Mathur Babu, Sambhu Malick and others.

"I have had many amazing visions. I had a vision of the Indivisible *Satchidananda*. Inside It I saw two groups with a fence between them. On the one side were Kedar, Chuni and other devotees who believe in the Personal God. On the other side was a luminous space like a heap of red brick dust. Inside it was seated Narendra (Swami Vivekananda) immersed in *Samadhi*.

"Therefore I feel that it is the Divine Mother Herself who dwells in this body and plays with the devotees. When I first had my exalted state of mind, my body would radiate light. My chest was always flushed. Then I said to Divine Mother: 'Mother do not reveal Thyself outwardly. Please go inside.' That is why my complexion is so dull now. If my body were still luminous, a crowd would always have thronged here. Now there is no outward manifestation." (*The Gospel of Sri Ramakrishna*, 1986, pp. 396, 813, 830, 831, 832, 934)

Immortal Bliss

The devotee who attains God enjoys eternal bliss, the true joy of life. This is another aim of *Bhakti*. Sri Ramakrishna describes God as the "Sea of Bliss." One of the beautiful names by which Nammalvar, the greatest among the *Alvars*, addressed God is "*Aravamudu*" which means "Nectar Endless." How is the devotee able to enjoy eternal joy? His intense love for God destroys desire and anger, the dreadful enemies of eternal bliss. In the *Bhagavad Gita*, Lord Krishna says:

> "With the self unattached to external contacts he discovers happiness
> In the Self; with the self engaged in the meditation of *Brahman*
> He attains to the endless happiness." (5. 21)
> "He, who is able, while still here (in this world) to withstand
> Before the liberation from the body the impulse born of desire and anger,
> (He) is a Yogi, he is a happy man." (5. 23)
> "When the perfectly controlled mind rests in the Self only,
> Free from longing for the objects of desire,
> Then it is said, 'he is united.'" (6. 18)

Without union with the Self (God), neither harmony nor balance is possible. Sage Sandilya, in his *Sandilya Sutras,* says, "Eternal happiness accrues to the devotees." (8) *Bhakti* is a blissful experience. Worldly pleasure and celestial joy pale into insignificance in comparison with the Divine Bliss. The *Bhaktirasayana,* a scripture on *Bhakti*, opens with a verse which states, "*Bhakti* is the supreme goal of man's endeavour; it is an incomparable and unalloyed bliss." The *Narayaniyam* of Narayana Bhatta (1588-1646), a great saint of Southern India, says, "Devotion to God, which is sweet, in the beginning, in the middle, and in the end, gives the highest bliss."

In *Tiruvacakam,* one of the most popular devotional books of South India, the great Saiva saint Manikkavacakar (eighth or ninth century A.D.) sings:

> "O inscrutable mystery, who eludes speech and thought,
> Indissoluble as the sweet juice of cane and cream
> Commingled in the fresh pure milk, Thy distilled bliss permeates
> In the heart of Thy beloved devotees."

Jesus says, "Enter thou into the joy of thy Lord." (St. Mathew, 25. 21)

Perfection

By attaining the Supreme Devotion, the devotee realizes his perfection, and immortality, and becomes contented. (*N.B.S.,* 4) When the intense love for God arises in the heart of a devotee, he becomes perfect, i.e., he realizes his Oneness with God or Divinity that is already within him. Jesus says, "Be ye, therefore, perfect, even as your Father... is perfect."

This perfection is not something to be attained, for, the true Being, the Self, is one with *Brahman*. Only ignorance veils the existence of God within us. Ignorance is the sense of ego, which arises through identification of the Self with the mind, senses and body. When the ego is dissolved, the natural perfection of the Self manifests itself, as the sun shines when the clouds clear away. The role of spiritual practice is only to remove this cloud of ignorance. The perfection is also liberation. When the bonds of ignorance are cut asunder, the devotee attains not only freedom from all imperfections and limitations but also from the cycle of births and deaths. Attaining immortality through knowledge of the Self or the indwelling God does not mean continuity of existence within time and space. It means realization that the Self is eternal and beyond time and space.

Contentment

When one reaches the highest perfection, one is contented with oneself. In the *Bhagavad Gita*, Sri Krishna says:

> "For one who rejoices only in the Self, who is satisfied in the Self,
> Who is content in the Self alone, verily there is nothing more to do." (3. 17)

The illumined sage does not depend on external objects for his happiness. He is quite satisfied and finds joy, bliss and contentment in the Self. He has complete satisfaction, which is a natural accompaniment of Realization. This is not one that is consciously aspired after. For such hankering is a kind of desire, and becomes an obstacle in the way of its achievement.

What should the devotee's attitude be before he can aspire to the highest? Kunthi, mother of the Pandavas, prays, "O Lord, let hardships and sorrows come always, if during their course we are blessed with Thy vision." (*Bhagavatam*, 1. 8. 24) Such should be the attitude of a real devotee. To realize God, a devotee must make persistent endeavour with an absolute indifference to pleasure or pain. His devotion should be a pure love that loves even in the midst of suffering. Madame Guyon, in her *Acquiescence of Pure Love*, says:

> To me it is equal whether love ordains my life or death, appoints pain or ease.
> My soul perceives no real ill in pain; in ease or health no real good she sees;
> One good she covets, and that good alone – to choose Thy will,
> From selfish bias free, and to prefer a cottage to a throne,
> And grief to comfort, if it pleases Thee.

When God is realized, it always brings to the devotee an unperturbed sense of satisfaction, which, "having obtained the devotee considers no other gain superior." (*Bhagavad Gita*, 6.22).

A discontented person, even if he is a billionaire, a prince or an emperor, is really a slave, for he has not been able to conquer his desire; but a contented person, whether he owns a cottage, or not, whether he possesses a piece of cloth to cover himself or not, reigns supreme over all circumstances. Such a person is a real master. In the *Bhagavad Gita*, Lord Krishna says:

> "The Yogi who is satisfied with the direct knowledge of the Self,
> Who has conquered the senses, and to whom a clod of earth,
> A piece of stone and gold are the same, is said to be harmonized." (6. 8)

On attaining perfection, the devotee becomes absolutely free from desire, grief and hatred; he does not rejoice over anything; he has nothing to gain by action. (*N.B.S.*, 5) This description is similar to the description of the sage of steady wisdom given by Lord Krishna in the following verses in the *Bhagavad Gita*:

> "When a person completely casts off, O Arjuna, all the desires of the mind
> And is satisfied in the Self by the Self,
> Then he is said to be one of steady wisdom." (2. 55)
> "He whose mind is not shaken by adversity, who does not hanker
> After pleasures, and who is free from attachment, fear and anger, is
> Called a sage of steady wisdom." (2. 56)
> "He who is everywhere without attachment on meeting with
> Anything good or bad, who neither rejoices nor hates,
> His wisdom is fixed." (2. 57)

The realized devotee has no attachment even to his life or body since he identifies himself with the supreme Self or God. He is quite indifferent to pleasure or pain, honour or dishonour as he is united with God.

Desirelessness

Desires arise from a sense of imperfection or limitation, which is characteristic of an egocentric person. Desires are like insatiable fire; the more fuel you put into it, the more lustily will it burn, and the more fuel it will want; no one can appease the hunger of fire. Our desires are insatiable like it. If we follow the dictates of the senses we shall never be able to satisfy them. The more a person wants this thing and that thing, the more is he a slave to the senses. Greediness leads to self-destruction. Tolstoy tells a beautiful story to illustrate how greediness destroys oneself:

A businessman bought many pieces of land. When he went to visit his friend, he told him how he had gotten all this land for a great price: thirteen thousand acres for only one thousand rubbles. The friend could not believe it. He said, "How did you manage that?" He said, "Well, I went to that place where there are very few people. They don't speak our language. I took gifts to them and they asked me what I wanted. I said, "I want land." They said, "Take as much land as you want." I gave them one thousand rubbles, and they gave me thirteen thousand acres!"

His friend was absolutely thrilled. He was always on the look out for more land, so with another man he set out for that place. When he got there, every thing was exactly the way it had been described. The people were very nice and sweet. He gave them gifts that he had brought with him – silk, tea and spices.

After the people had offered their hospitality, they asked, "What do you want from us?"

The man said, "I want your land."

They said, "Come here tomorrow morning. Whatever you can cover within a day is yours."

"The man was very surprised. He asked, "What does that mean?"

They replied, "Come tomorrow. We'll explain every thing then." All night he could not sleep, he was so excited. Before sunrise he got up and went there. All those people were waiting for him. They said, "Put your hat here." He put his hat down on the ground. They said, "Put one thousand rubbles in this hat." He did it. Then they gave him a shovel and said, "Take this shovel and start walking. Wherever you make a turn, dig up the earth a bit to mark that this is your land. Later on we'll go and measure the land and it will be yours. There's only one condition: You must be back here by sunset, or you'll get nothing."

What a bargain! He took a water bottle and a few pieces of bread, and he started walking toward the rising sun. It was an incredible feeling. He walked and walked. He began to walk further and further, faster and faster. He started to run. He kept on running and running.

Suddenly he realized it was four o'clock. He had to get back before sunset. So he turned around and began to run toward the place where people were gathered. He had to run very fast, because he had started at six O'clock in the morning and now he had so little time to retrace his steps. As he ran and ran, he kept asking himself, "Did I do too much? Will I be able to get back?"

He was running very fast. He had come very close to the place. In fact he began to hear the people cheering and cheering, and their voices gave him a little strength. He was running. He was sweating, his mouth was parched, his lips were breaking and cracking, his heart was hammering, and his legs were feeling weaker and weaker; yet the spirit was willing, so he kept running.

Just before sunset, he climbed the final hill and he saw the people. He arrived and did a full prostration. Blood began to flow out of his mouth. The people went to help him up, but there was no one living in that body any more. They took the shovel and dug up the earth, and buried him. The trench that they dug for his body was only six feet long and three feet wide.

What can keep us together regardless of what we possess or do not possess? The scriptures say it is love; it is devotion. If you do not have love, if you do not have devotion, then no matter how much you possess it is all in vain. But if you have devotion, you can love even with nothing. In the *Bhagavad Gita*, Lord Krishna says:

> "He who is established in unity and worships Me, the indweller in all beings,
> That Yogi abides in Me, whatever his mode of living may be." (6. 31)

Once the ego is transcended and perfection is attained, desires no longer exist and there is complete fulfilment. The feeling of imperfection exists when a person finds something other than himself. But the perfect devotee is not aware of anything other than God, his own higher Self. Hence selfish desires have no place in him.

True devotees would not pray for any worldly object. They would not even pray for liberation. "Those devotees," says Rupagosvamin in *Bhaktirasamrta Sindu,* "who are delighted in the service of the Lotus feet of Krishna, do not desire even Liberation." (1. 2-13) However, they continue to have a higher craving for worshipping the Lord, singing and hearing His glories and serving the world, the manifestation of the Lord. Even the greatest Incarnations like Buddha, Christ and Ramakrishna had such higher craving. Buddha gave up the bliss of *Nirvana* that he might be born again and again to serve the world. Christ had the desire to save the souls of sinners and bring the Kingdom of God to earth. Muhammad wanted to teach the Arabs the principles of religion and spirituality.

Freedom from Misery

The realized devotee is free from misery which includes all kinds of worldly pleasures whether *sattvic, rajasic* and *tamasic.* (*N.B.S.,* 5) Why are pleasures also included under misery? As Patanjali says in *Yoga Sutras* that to a wise person even pleasant experiences arising out of virtuous deeds are painful, because they bring pain either as consequence or as anticipation of loss of happiness or as fresh cravings arising from impressions of past happiness or as counter-action of *Gunas.* (2. 15) All these are based on ignorance and ego and must be included under the common name of misery. One can expect to be free from such misery only after the realization of the Truth. The realized person is beyond both pleasure and pain, and is beyond all misery. The absence of misery, however, does not signify that he is devoid of sympathy for the miseries of others. Sympathy is Divine. The devotee's sympathy for the sufferings of others is an expression of his Divinity.

Freedom from Hatred

This is another sign of a perfect devotee. It is the obstructed desire that reappears in the guise of hatred and anger. So when a devotee is free from any desire or pain, he cannot be subject to hatred. If one chances to bite one's tongue with one's teeth, with whom should one be angry for causing that pain? This is the attitude of a perfect devotee who has become one with all Creation. Moreover, to the devotee everything happens only by the Will of the Lord, and if, therefore, he hates anyone, it will be equal to hating God Himself. So the realized devotee is free from hatred, hypocrisy, and violence, and he possesses supreme forbearance, straight-forwardness and other virtues without any conscious effort; he has no more to practice them as a discipline. (*Naiskarmyasiddhi*, 4. 69) The *Sreyomarga* also affirms the same idea: "All that precede the attainment of realization are means that are brought about, but they are inherent in the case of a perfect person."

However, the righteous indignation against the evils of society or against the oppressors of humanity is not the same as hatred or anger arising out of obstructed selfish desire. Sri Krishna's indignation against Kamsa, Narakasura and other demons or Jesus Christ's against the moneychangers inside the temple whom he whipped out or Mohammed's against the oppressors of the devotees of the God is not a manifestation of anger, but out of righteousness and service.

REFERENCE

1. A text dealing with the life of Rama and harmonizing the ideals of *Jnana* and *Bhakti*.

6

Characteristics of Bhakti

The Nature of Bhakti

In the *Sivananda Lahari*, Sri Sankaracharya explains the nature of Bhakti in the beautiful verse: "*Ankolam ... Sa Bhaktiriyuchyte.*" In this verse, he compares a devotee's intense love for God to five well-known examples:

1. The ankola fruits fall to the ground and the seeds are liberated. They are instantly and powerfully attracted to the trunk of the mother tree. They move in the direction of the trunk and stick to it.
2. The needle flies to the magnet, attracted by an irresistible force.
3. A devout and chaste wife lives constantly on the thought of her husband and service to him.
4. A creeper restlessly searches for a tree to entwine itself, and once it has caught a tree it winds itself round it inextricably, as it were, with a 'great affection and love.' Even if the creeper is violently pulled away from the tree, the instant it is released, it will rush back to the tree and wind itself around it.
5. The river streaming towards the ocean surmounts all obstacles and flows continuously till it attains the ocean.

Such should be the devotion of a devotee to the Lord – constant, intense and powerful.

CHARACTERISTICS

The main characteristics of Divine Love (Bhakti) are described below:

Love for Love's sake

First, Divine Love is a love in which there is no jealousy, no struggle of ego, and no desire for material advantage or exclusive possession. A true devotee worships God, not because he desires anything from Him, but because he is dear to Him. God is his Beloved whom he loves for the sake of love. So long as we expect anything, we do not love truly. Only when we have begun to love for the sake of love do we get true devotion. There is no bargaining or shopping in love of God. A true devotee does not even seek liberation, though, in spite of himself, he becomes liberated.

Knows no Fear

Second, Divine Love knows no fear. It is not based on fear of punishment. If a baby plays in mud and gets itself dirty does the mother throw away the baby? No. She picks up the child, washes it, and then takes it on her lap. God is more than our own father or mother. We are His children. How can He punish us? If we undergo any suffering at all, it is not God's punishment; it is the effect of our own past bad actions.

A nineteenth century poet-saint of Maharashtra, India, in a *Bhajan* (Devotional Song), says:

> What fear could exist for one whose mind is dissolved in
> The bliss of Divine Consciousness?
> Fear arises only out of the feeling of individual existence
> Which comes from identifying with the body.
> It does not survive in the state of Divine Consciousness.

Knows no Rival

Third, Divine Love knows no rival. To a devotee, God is the one and only Beloved. In the words of Sri Ramakrishna, "an illumined soul realizes the bliss of God while he is absorbed in meditation, attaining oneness with the indivisible Impersonal Being, and he realizes the same bliss when he comes back to normal consciousness and sees the universe as a manifestation of that Being and as a Divine play."

Al Hallaj, a Sufi, having experienced the ultimate Truth, declared, "I am the Truth, I am He whom I love, and He whom I love is I." Dionysius says, "It is the nature of love to change a man into that which he loves."

In the *Bhagavatam,* we read how the Gopis, the milkmaids, became Krishnas, when they lost themselves in their absorbing love for Krishna.

Self-evident

Fourth, Divine Love is its own proof. Unlike any other object whose reality cannot easily be recognized, Divine Love does not require some other proof to recognize it; for it is self-evident. Love is directly felt and experienced in one's own heart. That itself is its own validity.

Peace and Bliss

Fifth, the very nature of Divine Love is peace and supreme Bliss. It is experienced when a devotee has attained his union with God. One may find peace and joy in human love, but it is not lasting. The peace and joy that a devotee experiences through Devotion to God is lasting and continuous, and it also grows in intensity. (*N.B.S.*, 59, 60)

No Grief

Sixth, the devotee does not grieve at any personal loss, for he has completely surrendered himself, everything he has, and even the rites and ceremonies which are enjoined by the scriptures. In fact, the devotee does not feel that he owns anything of this world. When pure love blossoms, the devotee becomes free from worldly ties and anxieties. No earthly condition can disturb his peace.

Even though the devotee has surrendered himself to the Lord, he can continue to perform action, giving up the fruits of action to the Lord. In fact, the great *Avatar*s like Sri Krishna, Buddha, Christ and Sri Ramakrishna were teaching and preaching for the good of mankind. Sri Krishna says in the *Bhagavad Gita*:

> "There is nothing in the three worlds, O Arjuna, that should be done by Me,
>   Nor is there anything that should be attained,
>   Yet I engage Myself in action." (3. 22)
> "For, should I not continue to work untiringly,
>   Mankind would follow My path, O Arjuna." (3. 23)
> "These worlds would perish , If I did not perform action. . . . (3. 24)

If the Lord were to remain inactive, the people also will become inactive. There would be nothing for them to do.

Bhakti: its own Fulfillment

Seventh, Bhakti cannot be used to fulfill any desire, being itself the check to all desires. When Supreme Love for God arises in the heart of a devotee, he no longer has any desire left in him for the objects or pleasures of this world. Once a devotee realizes his identity with Brahman, all his desires melt away. Here is an incident from the life of Swami Vivekananda, which illustrates that Bhakti itself annihilates all worldly desires:

Early in 1884, Narendra's (Swami Vivekananda's earlier name) father, a lawyer of the High Court had passed away suddenly. He had not been able to make provision for the family, which consequently faced grave financial difficulties. Some times the members of the family had to go without food. Narendra tried to find some job. He got some temporary job. It brought no real security to his mother and brothers. So now Narendra decided to request Sri Ramakrishna to pray to the Divine Mother to remove his poverty. Sri Ramakrishna asked him to pray to the Mother himself. When

Narendra entered the temple, he saw Her as a living Goddess. Narendra was overwhelmed. Unable to ask for petty worldly things, he prayed only for knowledge and devotion. His heart was filled with peace. The Universe completely disappeared from his consciousness and Mother alone remained.

When Narendra came back from the temple, Ramakrishna asked him if he had prayed for the removal of his family's poverty. Narendra was taken back. He had forgotten to do so. The Master sent him back to the temple. But Narendra, standing in Her presence, again forgot the purpose of his coming. Thrice he went to the temple at the bid of the Master, and thrice he returned, having forgotten in Her presence why he had gone there. Sri Ramakrishna, seeing his devotion, assured him that his family would not lack for simple food and clothing. (Source: Christopher Isherwood, *Ramakrishna and His Disciples,* New York: Simon and Schuster, 1965)

Devotion: Powerful

Eighth, true devotion has wonderful powers. Through devotion a devotee can even extract Divinity out from a stone. "The goal can be attained very easily through the power of faith and devotion, and never through the power of reasoning" says Sri Ramakrishna.

It is true devotion that brings God-vision to us. No one can reach Him through mere intellect; nor even by the practice of austerities. He can be attained by true and pure love, unselfish and one-pointed love. God is love and love is God. One can realize and feel this only with a sincere and pure heart.

The power of Devotion is amazing. In the *Bhagavad Gita*, Lord Krishna says to Arjuna;

> "He, who sees Me everywhere, and sees everything in Me
> Never becomes separated from Me, nor I from him." (6. 30)

Everything an Act of Worship

Ninth, the true devotee works for God. His eating, drinking, sleeping, moving, his every act becomes an act of worship. In the

Bhagavad Gita, Lord Krishna says,

> "Whatever you do, whatever you eat, whatever you offer in sacrifice,
> Whatever you give, whatever you practice as austerity,
> O Arjuna, do it as an offering unto Me.
> Thus you shall be freed from the bonds of actions yielding good and evil fruits;
> With the mind steadfast in the Yoga of renunciation, and liberated,
> You shall come unto Me." (9. 27, 28)

The true devotee looks upon his body as a temple in which God dwells, and whatever he does through his body becomes an offering to the Lord. The constant thought of God as dwelling within enables him to free himself from all physical bondage. Thus true devotion brings the devotee to a state where he lives with his Ideal in every moment of his life. Nothing else can exist for him except the Beloved One, and he lives in Him and Him alone. His whole heart goes to the Beloved and the devotee becomes lost in Him. This blessed state leads him to the ultimate goal – the true vision of God.

True Devotion comes like a flood

Tenth, when the whole-hearted and one-pointed love awakens, it comes like a flood and washes off everything – ignorance, narrowness, fear, doubt, selfishness, etc. The Ideal alone is left shining in the heart. Then it becomes easy to renounce everything that is earthly, because nothing except the Beloved has any value. He alone is Eternal, Permanent and Unchanging, and all other things are transitory and changing.

Renunciation becomes Easy

Eleventh, the true love for God enables one to renounce the things of the world without effort. Christ could resist the temptation of Satan and give up all earthly power because of His intense love for His Father in heaven. When we have whole-hearted devotion to God, nothing of this world can tempt us, and so renunciation becomes easy. This is the idea of "taking the burden" given by Christ. When we take refuge at the feet of the Lord or cast our burden on Him, we love Him, our whole heart goes to Him and we are no longer involved in the world. Then our burden drops off. This is true renunciation.

Humility

Twelfth, when real devotion comes to the devotee, he grows humble and all loving. Whatever he sees is Beloved, therefore he becomes the servant of all, and through every living creature, he serves God. Devotion to God implies humility, and humility attracts God's grace. Sri Chaitanya emphasizes humility in a prayer:

> "Be humbler than a blade of grass, be patient and forbearing like a tree,
> Take no honor to thyself, give honor to all,
> Chant unceasingly the Name of God."

Divine Love: open to All

Thirteenth, the Path of Divine Love is open to all. Faith in the grace of God is enough for the practice of the path of Bhakti. Even illiteracy is not an obstacle, nor a previous record of a vicious life. Lord Krishna in *Bhagavad Gita* declares:

> "Even if a hard-baked sinner worships Me with devotion,
> He, too, must be regarded as righteous, for he has rightly decided.
> Soon he becomes righteous and attains to eternal peace.
> O Arjuna, know thou for certain that My devotee never perishes." (9. 30, 31)
> "For those who take refuge and abide in Me, even those of sinful birth
> Attain to the Supreme Goal." (9. 32)

Sage Sandilya affirms, "Every one, to the lowest-born, is eligible to follow the path of devotion."

Untouchable saints like Nanda, Ravidas, Kannappa, Tiruppanalvar and Tirumangaialvar, and female devotees like Mirabai, Avvayar and Andal have graced this land to bear witness to the catholicity of the path of devotion. Moral wrecks like Ajamila, Ratnakara, Tondaradippodialvar, Narayana Bhatta and Vivamangal are shining examples for redeeming even the worst sinners. Saints like Kabir, Nanak, Tukaram and others prove that the path of Divine Love does not need book learning.

Bhakti – its Own Fruit

Last, but not least, another important characteristic of Devotion to God is that Bhakti is its own fruit or the goal itself. (*N.B.S.*, 26). That is, it has no cause and it is not the effect of anything else. Spiritual practices cause the destruction of ego and purify the heart. But *Para Bhakti* (Supreme Devotion) as spiritual experience is not mere destruction of ego. It is the eternally perfect nature of the Self revealing itself spontaneously when the obstructing causes are removed. Efforts do not induce this experience. They are powerless without the Grace of God. If a devotee takes one step towards God, God takes two steps towards the devotee. God comes to us in many forms, but we do not always recognize Him, unless He reveals Himself to us, and until we have deep devotion to Him.

7

The Philosophy of Bhakti Yoga

Meaning

Bhakti means devotion. The word *Bhakti* comes from the root *Bhaj* which means 'to serve' or 'be deeply interested in.' Bhakti is, therefore, an intense attachment to God or deep interest in God.

The innate nature of all beings is to love an external object and this love is experienced in the heart. Truly, the Absolute alone is existent. A human being is only an ego separated from it. Love for external things is an unconscious internal urge to become unified with everything. It is the forerunner of Experience. Love is the craving. Experience is the fulfillment of it. None can live without love for something. Senses always turn outward. The mind is the main sense of perception. It perceives through the five senses. The senses do not work when the mind does not.

Concentration

Mind takes interest in countless objects of the world. It is folly on our part to allow the mind to run randomly in all directions. Yogis have found that the mind that is focused on one point can do great things. It is the concentrated ray of the sun, passing through a lens, focused on a paper, that burns it. Likewise, the mind has to be concentrated on one object. It should not jump from one thing to another like a monkey. Such jumping is the way of the worldly life. It should be stopped by controlling the mind through concentration or one-pointedness. The thought waves of the mind should cease. In that inner silence , the Absoluteness is experience.

To live in the continuous experience of love, we need to keep our mind focused and not get distracted. Focused mind makes us greater and greater. It is through concentration that scientists and others have achieved colossal things, and the sages have achieved the most magnanimous thing – God-realization. It is up to us to choose what to focus on. Whatever that focus is, it will make one great.

Once a beggar went to a king to ask for alms. As he was telling the king his trials in life, the iron rod that he carried dug into the king's foot. The beggar was unaware that the king's foot was bleeding. After he had told the king all his problems and the king had given him some money, he left.

Afterward, while they were bandaging the king's foot, the minister asked him, "O king, why did you let him do this to you?"

The king said, "He came to me with his problems. He already had enough to bear. There was no need for me to tell him about the problem with my foot."

This shows a gracious heart. With most of us, if someone comes to tell us their problems, we say, "Listen to mine first." We behave in this way due to ignorance. We do not know what we are doing. But the king knew what he was doing: he was ruling his subjects and giving his life to them. He had a focus. Therefore, he had a selfless agenda; he could embrace all of them.

We can concentrate our minds on any thing – a sense-object, wealth or family member. However, this kind of concentration is not of the highest ideal. Love and attachment towards external, pleasurable objects binds and chains us to a cycle of innumerable births and deaths. But concentration on selfless love for God is a ladder to Liberation.

Emotions

Emotions are generally considered as a hindrance to Self-Realization. But only certain emotions are of a binding nature, while certain others will liberate us from bondage. Devotion to God does not rouse in us any binding emotion. It is pure emotion devoid of carnality and attachment. The love for God rouses the purest of emotions and it is far superior to the uncontrollable emotions that overpower us day and night. Those who cannot still all emotions can have at least pure emotions. This is the significance of Divine emotion in Bhakti Yoga. Love for God can never be the type of love cherished towards spouse, children and property. There is a fundamental difference between love for God and love for worldly objects. The former is unselfish, while the latter is based on selfish desires.

Liberation

How does love for God give us Liberation from the cycle of births and deaths? Man is an egoistic entity. Ego is his enemy. He feels that he is entirely different from other things of the world. He is convinced that he is sharply marked off from others by his physical body. He is sure that he is only the body. When he says "I," he always refers to his body. It is very difficult to separate the "I" from the notion of the body. None can easily get rid of this notion of the body as the real Self. The aim of all Yogas is to root out this sense of ego. And Bhakti Yoga is an easy method to destroy the sense of separateness or egoism. It annihilates the thought-waves of the mind and fills it with God-consciousness.

Divine Love

A devotee says, "O Lord! I am Thine. All is Thine. I am nothing. You are everything. O Lord! You are everywhere. I cannot even move, for You are everywhere. I am walking over Your Being. I am seeing You everywhere. You appear as human beings, animals, birds, trees, plants and so on. You have become everything. I have no separate existence. I am your instrument. I am doing nothing. You are doing everything through me. You are the Doer. You are the Enjoyer. I am nothing. Thy will be done."

Supreme Devotion

This is the highest form of Love. This is Divine Love. The ego cannot assert itself, for God alone is everywhere. The mind cannot modify into the modifications of sense-objects, for to the devotee, there is no object except God. Who is there to be loved or hated? The true devotee is therefore blissful at all times. His mind cannot think of anything. For, everything is God. Wherever the mind goes it experiences *Samadhi*, for it does not find an object of enjoyment. God is filling every bit of space. The saint and the sinner, the virtuous and the vicious, the good and bad, the man and the animal, all are forms of God. How can the mind deal with them in an unrighteous way? There the mind experiences *Samadhi*. *Samadhi* is thought-less Consciousness. This is Supreme Devotion. This is one with the Vedantic Realization. The effect of both is the annihilation of the ego or destruction of the mind. The mind cannot exist where there is no object of perception. God is pervading the entire universe. He is the earth and the heaven. He is the mother and father, and sister and brother. God is the consummation of all love and aspiration, desire and ambition. He is the Ideal to be attained.

No Objective Consciousness

Objective Consciousness is dead when the presence of God is felt everywhere. The sense-objects are transformed into Divine glory. Spouse is no more an object of lust, and money is no more an object to be coveted. All is God. All are to be worshipped. "The ass, the dog and the untouchable are to be worshipped" said Sri Krishna to Uddhava, His disciple. This is equal to the saying "*Sarvam Khalvidam Brahman*" (Everything is Brahman).

Annihilation of the mind and the ego

"Control the mind, annihilate the ego." This is the essence of all Yogas. This is also the ideal of Bhakti Yoga , and it is a very sweet and easy method. One need not curb his emotions and one need not run to the forest. He just has to direct his emotions to God and see Him present in the world. This is the essence of Bhakti Yoga. The *Upanishads* declare that Bhakti is a reflection of the love for the Self. Only the names are different. One calls it Self, another calls it God. Names do not matter much. It is the feeling that counts, and the feeling is the same.

Self-surrender

Self-surrender is the highest form of Devotion. It means surrender of the ego or individuality. What remains is the Absolute of the Vedantins. There is no difference between Vedanta and the highest form of Bhakti. A devotee surrenders the ego and a Vedantin disintegrates the ego. Anyhow the ego is not there in both. Their ideals are the same. Whether one eats rice or wheat, the purpose is the same – to appease hunger. And there is no conflict between the two.

Grades of Devotion

There are two grades of devotees. The lower category of devotee feels that everything is God except himself. He feels that he is the only one who is not God, and all else is God. This is lower Devotion and the presence of ego hampers the ultimate experience. The higher kind of devotee feels that he is also included in God and he has no separate existence. His ego is eliminated completely. This is the state of Supreme Devotion (*Para Bhakti*) (For details *see* chapter 10. Stages of Bhakti: 2. Supreme Devotion.) Here the emotions stop and the devotee becomes a calm ocean without waves. His mind is stilled and it merges into the Universal Truth. This is the culmination of Bhakti which supreme devotees like Chaitanya, the saint of Bengal (India) experienced.

Unselfish Love

Love for God should be unselfish. There should be no earthly motive behind Love for God. Otherwise, it becomes only a modification of infatuation and delusion. Those who seek redemption from distress, wealth and knowledge are all selfish devotees. They do not have the highest form of devotion. They are deluded by worldliness. A true selfless devotee is filled with emotionless peace. He loves God for love's sake. He does not seek anything from God. He just experiences God. When God is attained everything else is attained. The devotee is lost in the consciousness of God. He plunges into the ocean of Bliss. He becomes God.

Bhagavatam

The text of pure devotees is depicted in the ancient *Bhagavatam*, the cream of devotional Hindu literature.(omitted sentence) This text is pre-eminent for its philosophical depth, devotional exuberance and literary beauty. It is a treasure of Divine wisdom. The whole volume of eighteen thousand verses is completely saturated with the supreme expositions of Devotion, Renunciation and Wisdom. The Ideal of Renunciation and Knowledge of Rishabhadeva and Jadabharada, the supreme devotion of Dhruva, Prahlada and Ambarisha, the Wisdom of Narada, Kapila, and above all, the immortal life and Teachings of Bhagavan Sri Krishna to his devoted disciple Uddhava form the nucleus of the *Bhagavatam*.

Sri Krishna's teaching to Uddhava on the Philosophy of Love is:

Drink deep of My (God's) words, which are nectar. Study the lives and teachings of Divine Incarnations, the sons of God. Learn to find joy in the worship of the Lord. Sing His praises.

Being devoted to His service, worship Him with the whole soul. Ennobling also is the service to God's devotees. Learn to see God in all beings.

Let all your work be done as service unto the Lord. Let your every word extol His divine attributes. Free your mind from all selfish desires and offer it unto Him.

Renounce all enjoyments and pleasures; make sacrifices, offer gifts, chant the Lord's name, undertake vows and practice austerities. Do all these things for God's sake alone.

Thus by surrendering yourself unto the Lord through all your actions, and remembering Him constantly, you shall come to love Him. When you have come to love God, there will be nothing more for you to achieve. For when the mind is completely surrendered unto God, who is the Divine Self within, the heart becomes pure and tranquil, and one attains to Truth, knowledge, dispassion and divine power. Devoid of these is one whose mind is outgoing, seeking pleasure in objects of the senses.

Truth is love. Knowledge is seeing the oneness of the Self with God. Dispassion is non-attachment to objects of the senses, and Divine power is the control of nature, external and internal. (Source: *Bhagavatam*)

Devotion–the Spirit of every Religion

It is grave mistake to misrepresent and cavil at Divine Love, for the true spirit of every religion implies the adoration and love of God and the desire for Union with God. The highest conception of perpetual Bliss is not mere prostration and service, but a loving union with the Eternal. In emphasizing true Devotion as a method for Salvation, it is not meant that service and love of humanity should be denied. For, all is God, and he who serves humanity and other beings is serving God. He who loves his neighbour loves God. The devotee identifies himself with all the beings of the world. He feels the whole universe as a mere manifestation of God. Sincere devotees cannot go astray. They do not perish. Even the sinner is lifted up to the magnificent height of Emancipation. The kindness of God is immeasurable. God illumines the intellect of the devotees, and takes care of them at all times. The *Bhagavad Gita,* the *Bhagavatam,* the *Bible* and the scriptures of other religions bear witness to this fact. The devotees are guided by God and enlightened for the attainment of the Supreme Bliss and Peace. (Source: Swami Sivananda, *Essence of Bhakti*).

8

Worldly Love and Divine Love

Love

The human soul possesses the germ of an indescribable feeling called "love." Were it not for this spell, earthly existence would have been dull and dreary, and the human heart would have been dry like a desert. It is this feeling that creates friendship among persons, that binds the hearts of parents and children, and unites the souls of a husband and a wife. It is a mysterious glue that unites the hearts of all. Love is the greatest force on earth. It conquers the hearts of men and women. It subdues an enemy. In the words of Swami Sivananda, "Love can tame wild animals. Its power is infinite; its depth is unfathomable; its nature is ineffable; its glory is indescribable." Love is the crown of life. It is the greatest power. It is the true joy of every heart.

Robert J. Ingersoll, the great philosopher, describes love as follows:

"Love is the only bow on life's dark cloud. It is the morning and evening star. It shines upon the cradle of the babe and sheds its radiance upon the quiet tomb. It is the mother of art, inspirer of poet, patriot and philosopher. It was the first dream of immortality. It fills the world with melody, for music is the voice of love. It is the perfume of the wondrous flower – the heart. Without it we are less than beasts, but with it earth is heaven and we are gods in embryo."

There is no virtue higher than love. Love is Truth. Love is God. God is love. This world has come out of love, this world exists in love and this world ultimately dissolves in love. Live in love. Live to love. Live as love.

Expression of Love

The feeling of love expresses itself in various forms. Every living being loves itself. This love of self is found in animals as well as in human beings. Lower animals love themselves instinctively. Man alone is capable of knowing his love of self objectively.

Animal Love

In the lowest forms of animals, this love of self is confined to their own bodies. In the mammal species the love of self extends a little beyond their bodies. They also feel for their offspring and provide for them.

Animal love in man is confined to the love of his own physical body. Animal love forces one to seek pleasures of the senses and physical comfort. This love is very narrow and limited. It is egocentric. One who lives on this plane is extremely selfish, cruel and heartless towards others.

Human Love

The same love gradually develops and manifests itself in man as human love which is deeper and stronger in power. When animal love develops into human love, man feels for others as he feels for himself. Mutual attraction, mutual affection and attachment are the signs of human love.

True Love

True love rises above the physical plane. It is the attraction between two souls. It is for this reason that the *Upanishads* say, "A mother loves her child not for the sake of the child's physical form, but for the sake of the soul, the Atman, the Lord that lives in the child." The mother may not know it, but it is true. Similarly, "A wife loves her husband not for the sake of his physical form, but for the soul, the Atman that resides in him." Wherever there is true love there is the pure attraction between two souls. It is the stepping stone to Divine Love.

Ordinary human love binds the soul to earthly conditions and makes it attached to the pleasures of the senses. Governed by selfish motives and selfish desires, it blinds man from the truth. It brings sorrow, suffering and misery in the end. All evil and wickedness that is found in the world is the result of ill-directed feeling of selfish love, while all goodness and virtue are the results of actions proceeding from well-directed unselfish love.

A lover naturally seeks to be loved in return. This love is limited by personality and is confined within the narrow sphere of personal self. It strongly brings out the sense of "I, me, mine." Human love is subject to the influence of jealousy, hatred and egoism. It is far away from universal love. If any body speaks ill of you and uses harsh words, at once you are thrown out of balance. You get irritated, show an angry face, and retaliate. It is difficult to part with your possessions even though you see someone in need and in distress.

Divine Love

When love is directed towards God, it blossoms into Divine Love. Divine Love is the sweetest mystery of life. It is the greatest power on this earth. Divine Love is pure love. A true and sincere seeker loves God with his whole heart and soul. Love of God arises in a heart that is free from desire. Divine Love seeks no return. It leads to non-attachment to worldly pleasures and objects. It removes all fear of punishment. Fear and love cannot go together. A true lover of God does not fear anything in the Universe, for every thing in the Universe is a manifestation of God.

A great writer exhorts,
Love all God's creation, both the whole and every grain of sand. Love every leaf, every ray of light. Love the animals, love the plants, and love each separate thing. If you love each thing, you will perceive the mystery of God in all. And when once you perceive this, you will thence forward grow every day to a fuller understanding of it, until

you come at last to love the whole world, with a love that is all embracing and universal.

Divine Love brings a cessation of all sorrow, suffering and pain; it lifts the soul above all bondage, breaks the fetters of selfish attachment and worldliness. All selfishness vanishes and the soul enters into the abode of absolute freedom and everlasting happiness.

Real love knows no boundaries. It is love for the sake of love. A Sufi saint said, "You cannot say to your beloved, 'I love you just so much and no further.' When you truly love it is like giving a blank cheque of everything you possess, above all, yourself. . . . ."

A true lover of God loves everything in the world. Divine love opens his spiritual eye, and he realizes that everything comes from God, that everything belongs to God, and that all beings are His children. He sees all beings and objects of the world as God's manifestation. To repeat the words of Lord Krishna in the *Bhagavad Gita,*

"He who sees Me everywhere and sees everything in Me, of him I never lose hold and he shall never lose hold of Me." (6. 30)
"He, whose self is harmonized by Yoga, sees the Self abiding in all beings And all beings in the Self; everywhere he sees the same." (6. 29)

The true devotee feels that nothing in the world can happen without God's will. He surrenders his individual will to the Divine will. "All is Thine. Thy will be done." The true devotee cannot keep anything for himself. Even if he has just a little to keep life going, he will sacrifice that little to serve a needy one, and undergo deprivation and starvation willingly and with pleasure. He will rejoice that God has given him a rare opportunity to serve the destitute.

Divine Love brings the highest ecstatic or super-conscious state. The individual soul eternally communes with God, the Universal Spirit. Bhagavan Sri Ramakrishna and great saints like Sri Swami Satchidananda attained to this ecstasy and lived as perfect embodiments of Divine Love.

No Carnality in Divine Love

In earthly love, there is a need of sensual gratification, especially in sexual fascination (*kama*). But in Divine Love there is no carnality at all. In Love for God the mind naturally renounces sense pleasures. As Sankaracharya describes, when intense love for God arises in the heart of a devotee all his / her desires melt away; but in *kama*, mind is engrossed in sense pleasures. Therefore, Bhakti cannot be equated with *kama*. Modern psychologists would not be satisfied by the apparent difference in the object of interest in the two forms of love. They demand better reasons for giving spiritual experience a higher basis than man's instinctive energies that manifest as desires. A consistent attempt is made in modern times to find a sexual origin for the so-called higher experience of saints and mystics. Many a psychologist is inclined to attribute the spiritual intuitions of the saints and mystics to the suppression of basic instincts, especially, the sexual. No doubt various forms of sex taboos are prescribed for mystics and genuine spiritual aspirants. But the continence of a true aspirant and repression are poles apart in their method and result. Repression is accomplished through fear, unnatural application of force and ignorant evasion of the problem. Its result is that sex tendencies remain submerged, and manifest as fantasies accompanied by mental disorders. On the other hand, in the case of a genuine spiritual aspirant practicing continence, sex is fearlessly and intelligently faced and thus, instead of being allowed to remain submerged as a crude animal instinct, its energies are transformed into a higher power for the enrichment of man's psychic being. Continence itself is not spiritual illumination. Perfect continence provides the right subjective environment, a pure mind and body, where, spiritual experience can gain full-expression.

In fact spiritual experience is a special distinctive tendency different from other instincts. Prof. Rudolph Otto in his *Idea of the Holy*, defends a *priori* character of religious instinct and gives it the new name of *Numinous*. Starbuck, in his paper on the *Instinctive Basis of Religion*, maintains that the cosmo-aesthetic and teleo-aesthetic senses constitute the ultimate religious instinct in human nature. Prof. Hocking, in his *Human Nature and its Remaking,* speaks of the instinctive motive of religion as a specific craving for restoration of creative power. The saints possess special tendencies and aptitudes. The Hindus' attribute such endowments to tendencies acquired in previous births. The following verses of the *Bhagavad Gita* confirm this statement:

"There (born to a family of wise Yogis) he comes in touch with the knowledge
Acquired in his former body and strives more than before for perfection." (6.43)
"At the end of many births the wise person comes to Me,
Realizing that all this is the inner Self (Vasudeva);
Such a great soul (*Mahatma*) is very hard to find." (7.19)

The spiritual experience is something unique in itself and not attributable to any of the other instincts.

Worldly Love *vs.* Divine Love

In our daily life, we love our friends and family. It is true that sometimes we may go out of our way to help those whom we love. We may, for example, keep awake all night in order to attend to the needs of our sick parents. We may walk alone in the middle of the night to a doctor to get help for a sick friend. These, however, are rare examples of ideal human love. Human love rarely reaches the ideal of love for love's sake. More often than not, our human love is built upon selfishness. It is an external show of courtesy and manners. There is insincerity and artificiality. Pure love is natural and spontaneous.

Selfish human love leads to all sorts of vices like murder, theft and robbery. Most human beings are narcissistic. They become strongly attached to their physical form and devote their time to decorating their bodies.

Worldly love is relative and conditional. It has its demands, obligations and commitments. Pure love, on the other hand, is absolute and unconditional. To realize this true love, we need absolute wisdom. Wisdom and love substantiate each other.

Bhakti or Divine Love, as a means of God-realization, is also far removed from the crude notion of worship of spirits, gods, etc. out of fear or desire for favours. Love and fear cannot coexist. Fear proceeds from attachment to self, while true love is unselfish. Divine Love conquers all fear. So long as there is fear, it is not love. Love banishes all fear. Love never asks, never begs. Love loves for the sake of love itself. Even the idea of object vanishes. Love is the only form in which love is loved. This is the highest abstraction.

Love–an Inner Happening

Love is not awakened by saying, "I will give you love."

Swami Chidvilasananda [1] says, "It is an inner happening, beyond the reach of the physical senses. The eyes do not know, the nose does not know, the mouth, the ears, and the hands do not know. But something happens, and they are transformed. God's love for us does not depend on our faith and belief. It is an inner happening. Once we have experienced this inner happening, it is good to keep cultivating it so that we can keep this love alive within us all the time." (*Kindle My Heart – Vol. I*, 1989, p. 62)

True love is total absorption like a river merging into an ocean. It is so obvious, yet so hidden. Here is a story to illustrate this fact:

There was a man who crossed the Mexican border daily with a donkey carrying hay. The customs officer suspected that the man was smuggling something. He inspected the man, the donkey and the hay daily, but could never find anything suspicious. This went on for many years until both the man and the customs officer retired and became neighbours.

One day as the man was washing his donkey, the retired customs officer looked over the fence and asked him, "Friend, tell me, what were you smuggling all those years?" But the man was not willing to give out the information so easily.

The ex-officer coaxed him: "Look I won't tell anyone. It doesn't matter now. Both of us are retired. Please tell me. I am just curious to know what you were smuggling."

The man finally said, "All right, I was smuggling donkeys."

Love is so obvious, yet it is hidden, an inner experience. If we want it, it is there. If we do not look for it in the right place, many years may roll by and we will not find it. It is so obvious, yet so hidden.

Love of God – Love of Self

The *Chandogya Upanishad* shows that The Brahman to be worshipped and to be realized is the same as the Atman in the heart:

> "Verily, this whole world is *Brahman* ...Tranquil, one should meditate on it."
> "This is my Self within the heart, smaller than a grain of rice, ...
> This is my Self within the heart, greater than the earth, ...
> Greater than the sky, greater than these worlds."
> "This is the Spirit that is in my heart, this is Brahman." (3. 14. 1-4)

The *Bhagavad Gita* echoes the same truth:

> "I am seated in the heart of all, ...I am verily that which has to be
> Known by all the *Vedas*......and the knower of the *Vedas* am I." (15. 15)
> "The Yogis striving (for perfection) behold Him dwelling in the Self.....(15.11)

Love of God is nothing but Love of Reality or the higher Self of man, which he has forgotten; it is the same indwelling Divine Spirit. In the initial stages of spiritual practice, the Divine Spirit is conceived of as something different from the soul of man. Spiritual practice thus begins with separating imaginatively the God within from the empirical self and investing Him with all the noble qualities that one would like to develop in oneself. The individual gradually cultivates those qualities, and in course of time feels himself / herself as part of God and finally realizes Him as his / her own Self. This is echoed in the famous statement of Hanuman, the great devotee of Sri Rama. Once Sri Rama, seeing Hanuman among the sages assembled before him asked him, "Well, how do you look upon me?" Hanuman, the best of the Jnanis, seeing some great purpose behind this question, replied:

> "When I think of myself as an embodied being, then I am Thy servant;
> When I think of myself as an individual soul, I am part of You; but when
> I think of myself as the Atman, I am Thyself."

The Tests of Divine Love

How do you know that there is Divine Love or Devotion to God? According to Swami Vivekananda there are three tests for accessing this. The *first test* of Divine Love is that it knows *no bargain.* So when any one is praying to God: "Give me this and give me that," it is not love. Offering prayer for something in return is mere shopping.

Once there was a great king who went hunting in a forest, and there he happened to meet with a sage. He had some conversation with the sage and became so pleased with him that he asked him to accept a gift from him.

"No" said the sage, "I am perfectly satisfied with my condition; these trees give me enough fruits to eat; these beautiful pure streams supply me with all the water I need; and I sleep in this cave. I don't need anything."

The king said, "Just to purify me, to gratify me, pray take some present and come with me to the palace."

"All right. If that's going to make you happy, I'll come with you. Do you have everything you need?"

"Oh Yes, Swamiji, I am a wealthy king; I have everything."

When they reached the palace, the king wanted to show that he was also a great devotee. So he brought the sage to his beautiful shrine room, and said, "Swamiji, Please be seated. I will just finish my prayers."

The king performed an elaborate worship and prayed, "Lord, You have given me everything. But one thing is bothering me. The little country next to my kingdom is being ruled by another king. If I take over that country my position would be well secured. Lord, please grant me this boon."

When the king got up, he saw the sage going away. He ran after the sage, "Swamiji! Please wait. I have to offer my present."

The sage turned round to him and said, "I do not like to receive anything from beggars. You told me that you had everything. Yet you were begging for another country to be added to yours."

That is not the language of love. Love knows no bargaining; it always gives; total surrender to God.

The *second test* is that love knows no *fear*. Does the lamb love the lion, the mouse the cat? So long as one thinks of God as sitting above the clouds, with a reward in one hand, and punishment in the other, there can be no love. Love is never fearful. Imagine a young mother walking in the street and a dog is barking at her; she runs into the first house she sees. The next day she is in the street with her child and a lion jumps upon the child, what would she do? She would not run but would tackle the lion to protect the child. Love conquers all fear. The children of God never see Him as a giver of punishment or reward, but see Him as a loving father.

The *third test* is that love always commits to the *highest ideal*. When one has passed through the first two stages – when he has thrown off all shopping and cast off all fear – he begins to understand the true meaning of love. How many of us have very ordinary looking brothers or sisters? Yet the very idea of their being our brothers or sisters makes them beautiful to us.

The philosophy behind this is that each one projects his own ideal and worships that. All that we see is a projection of our own minds. The wicked see this world as a perfect hell, and the good sees it as a perfect heaven; the perfect person sees nothing but God. This universe is a manifestation of God. Love is the generating power behind all attraction, all sacrifices and all good and bad actions. Unattached, yet shining in everything is pure love and this love is God. (Source: Swami Vivekananda, *Religion of Love*, 1970)

9

Stages of Bhakti: 1. Preparatory Devotion
(*Apara Bhakti*)

Stages of Bhakti

Divine Love or Bhakti has a pre-mature stage and a mature stage of subjective experience. The premature stage is described as Preparatory Devotion or *Apara Bhakti*. The mature stage is known as Supreme Devotion or *Para Bhakti*.

This is how Sri Ramakrishna describes the degrees of Bhakti:

> "Bhakti matured becomes *bhava* (emotion). Next is *Mahabhava*. It is a Divine ecstasy; it shakes the body and mind to their very foundation. Then *Prema* and last of all is the attainment of God. *Prema* means ecstatic love of God. It is also known as *Raga Bhakti*, pure love of God – a love that seeks God alone and not any worldly end. Prahlada had it. It is love of God for its own sake. Chaitanya experienced the States of *Mahabhava* and *Prema*. When *Prema* is awakened, a devotee completely forgets his body, which everyone loves so intensely. He has no feeling of my-ness toward the body. Only *Isvarakotis*, such as Divine Incarnations, experience *Prema*. *Prema* is like a cord; by *Prema* God is bound to the devotee." (Source: *The Gospel of Sri Ramakrishna*)

According to Sri Ramakrishna there is also another kind of love, known as *Urjhita Bhakti*, an ecstatic love of God that overflows When it is awakened, the devotee laughs and weeps and dances and sings. Chaitanya deva is an example of this love. Rama said to

Lakshmana, "Brother, if anywhere you see the manifestation of *Urjhita Bhakti,* know for certain that I am there." (*The Gospel of Sri Ramakrishna,* p. 698)

Preparatory Devotion (*Apara Bhakti*)

Apara Bhakti is a preparatory discipline that preceeds the stage of maturation towards Divine Devotion. It is a disciplinary stage of devotion that involves effort and gradual achievement. It is a process.

Apara Bhakti is divided into *Gauna Bhakti* and *Mukhya Bhakti*. They are styled as *Vaidhi Bhakti* and *Raga Bhakti* respectively by Sri Ramakrishna.

Gauna Bhakti

Gauna Bhakti refers to a preparatory stage of discipline. It is associated with the *Gunas* or qualities of mind. Therefore, it is called *Gauna Bhakti*. It is classified in two ways. (*N.B.S.,* 56).

The first Classification

This is based upon the *Gunas* and dispositions of the devotees. Preparatory Bhakti is of three kinds: *Sattvic, Rajasic* and *Tamasic*.

Sattvic Bhakti. In this type of devotion, God is loved for His own sake; the goal and means are perceived by the devotee clearly and intelligently, and enthusiastic effort is made by him to realize the goal in the midst of all obstruction. He aims at freedom from the bondage of worldliness, and at attaining God. His love for God is pure.

A *Sattvic* devotee makes no outward display. He loves privacy. He meditates on God in absolute secret. He lives on simple food; there is no luxury in clothes, and no display of extravagance. He is very gentle, quiet, kind and humble.

Rajasic Bhakti This devotion is associated with extremely selfish desires and worldly ambitions. *Rajasic* devotee worships God for success in worldly ambitions, health and prosperity. He makes a display of his devotion before others. He puts a tilak on his forehead, a gold necklace of rudraksha beads around his neck and at worship wears a silk cloth and thus puts up a show and indulges in extravagance. He is engaged in restless activities to gain wealth, power and name for himself. He seeks God only as a means to gain his own selfish ends. The demons, Taraka, Hiranyakasipu, Ravana and others described in the *Puranas* are of this type.

Tamasic Bhakti. In this type of devotion, the devotee does not know clearly either the means or the goal. He is lazy, indolent, and prone to rely too much upon mere habit, custom and tradition.

Meaningless slavery, dependence on priests, fanaticism, faith in the magical power of incantations, fear of evil powers, resorting to occult methods in injuring others – these are among the signs of a *Tamasic* devotee. He literally extorts boon from God, just as a robber falls upon a person and plunders his money. His way is the way of dacoits.

The *Tamasic*, striving hard through his devotion, will reach the stage of *Rajasic* type. The *Rajasic* devotee can, through intensive devotion, reach the *Sattvic* stage. The *Sattvic* seeker can easily go to the higher stage.

Sri Ramakrishna classifies devotees into three categories. Devotees of the highest class see the Beloved Lord everywhere. They see God in Brahma, the creator of the universe as well as in a blade of grass. Devotees of the middle category see God within the shrine of their own hearts and know Him to be the inner Ruler. Devotees of the lowest group look toward the sky and say, "God is up there."

The Second Classification

This is based on the different motives that impel the devotees. These devotees are classified into three types, namely, *Arta, Jijnasu* and *Artharthi.*

Arta Bhakti. The underlying motive of this devotion is to rid one of the misery of birth and death, and to experience infinite bliss. *Arta* is discontented with all that finite material life can give. He is disgusted with everything, even with intellectual enquiry, because he is not satisfied with the book-knowledge of the All-pervading. He wants to experience It. He is detached from the objects of the world. He is totally under the influence of *Sattva*. He seeks God-realization itself.

Jijnasu Bhakti. In this devotion, the underlying motive is to get the highest knowledge of Reality behind the phenomenal appearances. *Jijnasu* wants to venture into the daring realms of investigation. Intellectually alert and vigilant, he lacks deep spiritual insight. He seeks to understand the Higher Reality. Being intellectually discontent, he becomes adventurous and aggressive, prompted by *the Rajasic* quality in him. He studies, inquires, discusses and contemplates to gain the knowledge of Self. Learning and hearing about God, this seeker of knowledge strives to ascertain the Glory of God.

Artharthi Bhakti. Here, the devotee wishes to see the establishment of the Divine kingdom on earth. He is dissatisfied with the prevailing conditions of the world. He finds that the world is plagued with exploitation, injustice, deceit, atrocities and squalor in the midst of plenty. He finds that righteousness has declined and unrighteousness has risen. He fervently prays for the establishment of righteousness on earth.

Of these classes of Bhakti, the second is superior to the third, and the first is superior to the second. (*N.B.S.*, 57)

Bhakti with Motive

All the above classes of Bhakti have a motive behind the devotion for God. Any devotion with a motive, whether lofty or otherwise, is not pure devotion.

Practices

The *Apara Bhakti* involves the practice of Bhakti Yoga or practices meant for the cultivation and purification of the various faculties of mind and their harmonious coordination. Bhakti Yoga is mainly concerned with the refinement and purification of the emotions.

The development of emotions is achieved through loving God or Devotion. Various disciplines such as non-attachment, cultivation of virtues, prayer, etc. are practiced to achieve this devotion. (For details, *see* chapters 19 to 26 Cultivation of Bhakti).

All spiritual practices, both positive and negative, as pointed by Sankaracharya in his commentary on verse 2. 55 of the *Bhagavad Gita* are meant for developing in the aspirant the qualities and characteristics required for the attainment of the goal. Renunciation of ego, detachment from worldly objects and sensual pleasures, cultivation of good virtues and devotion to God form the cornerstone of all spiritual disciplines. (*Bhagavatam,* 4. 22. 21). Real success comes only when the positive and negative practices are pursued simultaneously. Negative practices prevent an aspirant from retrograding and positive practices lead him toward Divinity. In the early stages, objects that cause attachment and bondage should be avoided. Practice of renunciation of sense objects must, however, be voluntary and not forced.

One has to avoid scrupulously objects, thoughts, feelings, and actions incited by self-interest. One should not stop with this. One should follow it up by positive practices. One should actively try to express in one's life all the Divine virtues, which one inherently possesses. The cultivation of positive good thoughts, feelings and ambitions, ceaseless effort to lead a virtuous life, performance of selfless service, seeking the Knowledge of Self, self-surrender to God – all these are needed for the spiritual development of a human being. The cultivation of *Sattva* and the acquisition of Divine virtues are spoken of in chapters 14 and 16 of the *Bhagavad Gita.*

Mukhya Bhakti (Primary Devotion)

This is the advanced stage of *Apara Bhakti*. This is also known as *Raga Bhakti* (Increased attachment to God) and *Ekanta* (one-pointed) Devotion. This stage prepares the mind of the devotee fully and perfectly for the final stage of Divine Devotion.

The moment conditions are favorable and the mind becomes sufficiently pure as a result of the spiritual disciplines and practices, the experience of *Mukhya Bhakti* comes automatically to the devotee by the Grace of the Lord . (*N.B.S.*, 53). This devotion is an ecstatic or pure love of God – a love that seeks God alone and not any worldly end. Prahlada had it. It is love of God for its own sake. When this love awakens, the devotee not only feels the world to be unreal, but also forgets the body. He has also no feeling of my-ness toward the body. Only the Divine Incarnations express this pure love.

Narada proclaims to the world the reality of the experience of such devotion. Psychologists like Leuba may often be tempted to relegate such experience to the category of hallucination and hysteria. However, without a purified mind, it is not possible for anyone to equate with this experience, moreover, describe it! Pure love is a subjective experience (*anubhava*) which cannot be observed by another person and hence it defies description by anybody except the subject who experiences it.

The Nature of Mukhya Bhakti

Beyond Gunas. Mukhya Bhakti is beyond the three *Gunas: Sattva, Rajas* and *Tamas*. It is untouched by the *Gunas* as the sky is untouched by the drifting clouds.

Devoid of Desires. Mukhya Bhakti is devoid of desires. The desires for the worldly objects are refractions of Divine Love caused by the prism of egoistic impure mind. When True Love is tasted, all those desires disappear. How can the stars continue to twinkle when the sun has risen?

Continuous Growth. The experience of joy derived from worldly objects is neither permanent, nor continuous. Repeated experience of the same object leads to a sense of boredom. Divine Love is different from this trend. It is boundless in its novelty, limitless in its expansion, and continuous in its growth. Delight leads to increasing delight, love leads to increasing love and the devotee's life is ever blissful.

Uninterrupted Flow. Divine Love flows in a continuous thread like the flow of water in a perennial stream. It is not interrupted by the distractions presented by the senses and worldly desires.

Subtlety. Divine Love is subtler than the subtlest object of the world. It cannot be grasped adequately even by the subtlest intellect. It manifests only in the purified mind just as the moon reveals itself in a placid lake.

Experience itself. Divine Love is not an object of experience, but it is an experience itself. In this experience, the illusion of the world, the veil of *Maya* (Cosmic Illusion) and the erroneous sense of separation from Divinity are destroyed. The devotee discovers himself to be the eternal ever-shining Consciousness. This Self is Divine Love.

Everything Divine Self. When the mind is dominated by ignorance and the consequent sense of ego, the world appears to be other than the Divine Self. The prism of the mind refracts into the multicolored objects of the world. But when the devotee attains the Divine Love through the practice of Bhakti Yoga disciplines, the shining light of Divine Love removes the darkness of ignorance; then he sees the Divine Spirit underlying all names and forms. He sees his Divine Beloved everywhere and in everything. He sees the beauty of the Lord in the sky, His laughter in the thundering clouds, His Bliss in the surging ocean, the melody of His flute in the Soft breeze, His face in the shining sun, His smile in the gentle moon, His exhilarating touch behind every experience of pleasure and pain, and His hand behind all the happenings of the world.

The devotee is immersed in the awareness of the Divine Spirit. In his thoughts, words and hearing perception he is aware of nothing but that Divine Love. (*N.B.S.*, 55). This blissful experience is described in a verse in the *Vedas* as "Sweet like honey, the oceans surge, Divine nectar ascends through the herbs and trees, and it is this very nectar that permeates even the very dust of the earth."

Chandogya Upanishad says,

"Where one sees nothing else, hears nothing else, and knows nothing else, that is the state of Infinite Bliss (*Bhuma*). But where one sees something else,

hears something else, and knows something else, that is the state of finite world of the mind and senses (*Alpam*). In that which is Infinite, there lies immortality, but in that which is finite, there lies (the fear of) death." (7. 24. 1)

Peace and Bliss. Devotion to God is of the nature of peace and bliss. (*N.B.S.*, 60) The devotee is led from sweetness to greater sweetness, from peace to greater peace and from joy to spiritual bliss. Therefore, the disciplines of Bhakti Yoga are characterized by spontaneity, inspiration, peace and joy.

Trust in the Divine Plan. A devotee performs actions that promote harmony in the world, but he is not affected by the enormous sufferings of humanity because he develops a childlike trust in the Divine Plan. (*N.B.S.*, 61). The human sufferings arise out of karmic effects. Bad actions lead to suffering and virtuous actions to happiness. An enlightened devotee is beyond virtue and vice beyond pleasure and pain due to his complete self-surrender to God.

Selfless Service. An enlightened devotee is free from desires, and he has no need to perform actions for fulfilling desires. However, he is not an idler. He continues to perform actions without egoistic involvement for the uplift of the world. (*N.B.S.*, 62). He works as an instrument of God and offers the fruits of actions to the Lotus Feet of the Lord. Such actions bring about purity of the heart and thus enable the devotee to experience the increasing sweetness of Divine Love.

Incessant Devotion. The *Mukhya* Devotee rises above the three kinds of devotion with motives–*Arta, Jijnasu* and *Artarthi*, and also *Sattvic, Rajasic* and *Tamasic* devotions. He begins to experience the Ecstasy of Divine Love (*Mahabhava*), and becomes incessantly devoted to God. (*N.B.S.*, 66).This is one-pointed (*Ekanta*) Devotion.

A one-pointed devotee constantly manifests love and love alone like a devoted servant or a devoted wife. A devoted servant finds his happiness in the happiness of his master, and he is ever self-effacing and sacrificing in nature. Likewise a loving wife has total dedication to her husband and lives for his happiness. The one-pointed devotee has developed total devotion to the Lord. He becomes fully immersed in Divine Love. He does not live for anything other than Divine Devotion. Such a devotee is supremely blessed.

10

Stages of Bhakti: 2. Supreme Devotion (Para Bhakti)

Para Bhakti (Supreme Devotion)

This is the final stage of devotion and the highest manifestation of love for God. The devotee feels totally one with the Supreme Lord. He merges himself entirely with the Lord and is absorbed in His sweet remembrance. Devi Bhagavata defines it as follows: "As oil poured from one vessel to another falls in an unbroken line, so when the mind in an unbroken stream thinks of the lord, we have what is called Para Bhakti or Supreme Love." Nammalvar, a reputed Saint of South India, experienced this Para Bhakti. He would caress the earth and say, "Oh! This is the Lord's earth." He would point to the sky and exclaim, "Ah! There is Vaikunta or heaven." He would feel the cool breeze and say, "This is my Beloved!" He would raise his hands toward the ocean and cry, "Behold the sea where my Lord reposes!" He would point to the hills and say, "See my Vishnu has come." When he saw a black mass of clouds he would say, "That is my Lord Krishna." Namdev, a great poet-saint of Maharashtra (India), was a Para Bhakta. Once, a dog ran away with his piece of dry bread. Namdev followed it, not with a stick to beat it, but with a cup of ghee to soften the bread lest it should hurt the dog's throat! What compassionate vision he had!

The Nature of Para Bhakti Chapter twelve of the Bhagavad Gita gives an exposition of Para Bhakti. Sri Krishna says:

"Those who, fixing their minds on Me, worship Me, ever steadfast and Endowed with supreme faith, are the Supreme Yogis." (12. 2)

It is this state of mind coupled with firm faith that makes Supreme Devotion. The devotee fixes his mind exclusively on God. He has no other thoughts except those of God. He is totally free from attachment and desires. He lives in God alone.

In the verses 13 to 19 of Chapter twelve, Lord Krishna describes the nature of the Supreme Devotee. He is established in God. He bears no ill will toward any one. He is compassionate to all. He has no sense of 'mine.' He is balanced in pleasure and pain. He is neither attached to pleasant things, nor hates unpleasant things. He does not hate even those who give him pain. He is as forgiving as the earth. The Supreme Devotee is ever content. Like the ocean that is ever full, his heart is ever full, as he has no cravings. He has controlled all the senses. He is endowed with a clear concept of God. He constantly meditates on God.

The devotee never injures any creature in thought, word or deed. He feels that the world is his own Self. He is unafraid of the world and does not hurt anything. He is free from all mental modifications – joy, envy, fear and anxiety. He does not rejoice when he gets desirable objects, nor he is unhappy when he gets undesirable things.

He is free from wants. He is pure in heart. He has renounced both good and bad. He possesses equanimity. He is the same to friend and foe, as well as in honour and dishonour. He is neither elated by praise nor pained by censure. He regards the whole world as his dwelling place.

Para Bhakti Inexpressible: "The real nature of the Supreme Love is inexpressible," says Narada. (N.B.S., 51) It is like a dumb man trying to express his experience of a delightful taste. The dumb man's joy can be only felt within. Likewise, the experience of God's love can only be felt within.

The disciples of Sri Ramakrishna had once urged him to describe the Supreme experience. When he made an attempt, he immediately went into Samadhi. Every time he tried to express it in words, he had the experience itself. Sri Ramakrishna used to say that the experience is like using a salt doll to measure the depth of the ocean.

As soon as the salt doll enters the ocean, it dissolves.

Manifests in Great Souls: Though Supreme Love is inexpressible, it is manifested in many great souls who have attained it. Their exemplary lives are guideposts to spiritual aspirants. While in the company of these souls, aspirants can feel their love flowing toward all beings and witness their blissful state of consciousness.

Atman's Real Nature: Supreme Love is the real nature of the Atman or Brahman (Brahman being love itself). In other words, Supreme Love is God Himself, not an attribute of God.

Nectar: Supreme Love is nectar, or Amrita (N.B.S., 3). God is immortal, eternal and supreme bliss. One who has tasted the nectar of Para Bhakti becomes free from the cycle of birth and death. Taittriya Upanishad describes God as rasa (nectar of sweetness):

> "Indeed that Brahman is Nectar. It is by attaining this Nectar That one becomes joyous. If this bliss did not exist in the heart, Who would have breathed, who would have lived? It is this Nectar that is the giver of delight."

Free from Lust: Supreme Devotion is bereft of any trace of lust because it is the nature of renunciation. A Supreme Devotee is completely free from the illusion of worldly desires, as the purifying water of Divine Love washes off the very roots of worldly desires and lust, and transforms them into Divine sentiments.

In the Bhagavad Gita Lord Krishna states that there are three gates to hell: lust, anger and greed (16. 21). Lust modifies itself into anger and greed, and so it is considered to be the greatest enemy of spiritual seekers. This enemy resides in the senses and mind. That is why control of senses and mind is an important discipline on the spiritual path.

A devotee detaches his sentiments from worldly objects and directs them to the Divine Self. He considers all objects as Divine possessions. He also invokes the Divine presence in every object, action and movement of life. He attains the highest state of renunciation when his ego-sense is completely dissolved in the experience of Divine Love.

Consecration of Actions: The cessation of desires does not mean fighting with desires, suppressing them or stopping all actions by taking a recourse to a life of inactivity. It is impossible for anyone to remain inactive even for a second. When devotee performs actions as an offering to God, he spiritualizes or consecrates all his actions (N.B.S., 8). Lord Krishna says in the Gita:

"Having renounced attachments, one who performs actions
As an offering to the Lord is not tainted by sin,
As a lotus leaf by water. Yogis perform actions
By body, mind, intellect and senses, renouncing attachment
For the purification of the Self." (5. 10, 11)

The consecration of actions is intimately connected with the spiritual disciplines of dispassion, renunciation of egoism, surrender to God and dedicating oneself to God.

Single-minded Devotion: Just as a devoted wife constantly thinks of her husband, or a miser thinks of his gold, a Supreme Devotee constantly thinks of God only. There is nothing to distract his mind. Wherever his mind goes, he sees the presence of God. With a vision steeped in Divine Love, there is no room for any other support in his heart (N.B.S., 10). By this one-pointed devotion to God, all his actions are easily renounced.

Selfless Actions: The enlightened devotee becomes a flute in the hands of the Lord – free from vanity and egocentricity, and devoid of selfish interests in life. All his actions conform to Divine will (N.B.S., 11). Whether he performs religious rites or service to humanity, lives the life of a recluse, or rules a country, he is ever established in communion with God. He is ever devoted to the good of all beings. He continues effortlessly to perform actions for the good of humanity without any sense of doership. Such actions flow through him spontaneously, for he is an instrument in the hands of God.

The realized devotee does not create any new karmas (effects of actions), because his actions are the outcome of Divine Will operating through him. He is not attached to actions or their results.

Beholding God in All Beings: God pervades the entire universe. Beholding Him in all beings and loving them all is the true adoration of the Lord. In this vision of love, good and evil, likes and dislikes have no significance. Love is based upon equality and a consciousness of universality. Supreme Divine Love is absolutely pure and flows from the heart of the devotee, flooding the world, just as the light from the sun shines equally on all. This love expresses itself in cheerfulness, contentment, self-sacrifice, forgiveness, compassion and peace.

Equanimity and Harmony: Supreme Love is founded on equanimity, bringing forth harmonious existence among all beings. The pure and dazzling power of the Spirit dwells in and pervades all beings and things. This Love is infinite and eternal.

While this love is beyond the implications of names and forms – it still works through them. It is beyond the sense of duality – still it reveals in multifarious ways. It is the omnipotent power that guides and controls all things. Supreme Love is not coloured by the conflict of opposites and the Gunas. It is spontaneous in expression. Its manifestation is based upon its indivisible unity and oneness with all that exists.

Supreme Love–Spiritual Realization

Narada says that the Supreme Love is identical with Spiritual Realization that is its own fruit (N.B.S., 26). The unfoldment of Divinity within is not the effect of any other cause. Spiritual Realization is eternal and infinite. One may then question the need for spiritual disciplines and teachings. God is already dwelling within, but He is covered by the cloud of ignorance, the veil of Maya (Illusion). Spiritual disciplines and teachings remove this veil of ignorance, thus revealing the Divinity within.

The Supreme Devotee finds God in all places, both outside the temples and churches as well as within. He finds God in man's wickedness as well as his saintliness, because God is already seated in his own heart, shining the eternal Light of Love. The spiritual aspirant who aims for the highest realization must accept Supreme Love as the highest goal. In the Chandogya Upanishad we read:

"One who knows, meditates upon, and realizes the truth of the Self, Delights in the Self, revels in the Self, rejoices in the Self." (7. 25. 2)

Shankaracharya says, "The illumined seer's delight is the Atman in joy and freedom." Nothing remains to bind him or fetter his freedom. He has transcended all limitations and the cycle of death and rebirth.

Universal Love

Supreme Love works with such an infinite vision that it breaks down all barriers. The devotee beholds the eternal Beloved of his heart manifesting in all beings and creatures, and envelops the entire creation. The intellect, detached and merged in the Super-conscious and static existence behind the manifestation, stands as an eternal witness, calm, serene and filled with radiance, supporting the workings of universal love and service for which the heart and body are the instruments.

The Para Bhakta forgets himself altogether and does not feel that anything belongs to him; then he acquires the state of His-ness. Everything – the whole universe – is sacred to him, because it is all His. He worships God in everything. He accepts all paths and all spiritual practices as equally valid as his own. He sees every one as part of his family. He sees all as equals in the eyes of the Lord, irrespective of external differences such as sex, color, caste or age. He serves all beings.

Glory of Para Bhakti

Superiority of Para Bhakti: Supreme Devotion, in comparison, is greater than the Yoga of Action, the Yoga of Wisdom and the Yoga of Meditation (N.B.S., 25). There are four aspects of human personality: Action, Emotion, Will and Reason. Corresponding to these four aspects there are four kinds of Yoga: Karma Yoga for purifying and sublimating the active aspect of one's personality; Bhakti Yoga (Secondary Devotion) for culturing and integrating one's emotions; Raja Yoga and Jnana Yoga for cultivating and transforming subtle intellect into discrimination and intuition.

Since aspirants are at different levels of evolution and possess different tastes and temperaments, each of them chooses one of the four aspects of Yoga as his path to spiritual advancement. But as a Yogi advances, the elements of other aspects of Yoga begin to blend in his personality. In the state of perfection, all four aspects are blended and transcended.

Sage Narada believes that the Path of Devotion, even in its secondary state, is more potent than the other Paths of Wisdom, Action and Meditation. A devotee seeks an increasing intensity of devotion and in this process he attains humility, self-effacement and surrender –indispensable attributes for true spiritual evolution.

Culturing and integrating one's emotions are more directly related to life's deepest experiences. When following any of the other paths, an awakening of sublime feelings produces an integrated personality. Therefore, even at the practice level, the Path of Devotion excels other Paths of Yoga. Love of God removes obstacles in the path, and brings about concentration of mind; it intensifies spiritual aspiration and fortifies the aspirant against the temptations of the world. It enables him to gain spiritual strength through Divine Grace, and secures success to him on the spiritual path.

Without devotion, those who follow the other aspects of Yoga may easily fall a prey to egoistic temptations. A Karma Yogi, instead of purifying his heart, may develop a pride in being a benefactor of the world. A Raja Yogi may be tempted to seek various psychic powers and thus drift away from the goal. A Jnana Yogi may misinterpret the great teaching, "I am Brahman," and develop a highly inflated ego. But with the development of greater devotion, these dangers can be averted. This is why the Path of Devotion is given greater importance by sages and saints and in religions.

End in Itself: The supremacy of Para Bhakti over other forms of Yoga is also because its means is the same as its end. (N.B.S., 26). Karma Yoga is the means to purifying the mind. Purification of the mind is also the means to focused Meditation. Jnana Yoga is the means for unfoldment of intuition and removal of ignorance. But in the advanced stage of Devotion, a devotee loves God for the sake of love itself. Supreme Devotion is God-realization. So it is the goal of all forms of Yoga.

Ego Checks: As explained previously, in every form of Yoga other than the Path of Devotion, one runs the risk of falling a prey to the influence of ego. In the Path of devotion, humility, self-effacement and surrender to God are the basic constituents of practice (sadhana), and therefore, devotion prevents ego from asserting itself (N.B.S., 27). This is beautifully expressed in the following poem of Saint Chaitanya:

"Being humbler than a blade of grass, more enduring than a tree,
Giving honour to those who lack honour, the devotee continues to be
Immersed in singing the glory of the Lord.

A devotee who enjoys Bhakti Rasa (the sentiment of Devotion) manifests eight expressions: choking voice, stammering, trembling of body, tears, insensibility, change in the facial expression, horripilation and swoon. The tears that flow from the eyes of the devotee are tears of immense joy. His whole body is thrilled with the experience of Divine Love.

Purifies the World: Indeed such a devotee is supremely blessed. He ever delights in talking about the Glory of the Lord. He is ever immersed in enjoying the nectar of devotion. No one can render a greater service to his family, his forefathers, his society and the whole world than a devotee who is immersed in Supreme Devotion. (N.B.S., 68). He radiates peace and harmony to all. Sutasamhita says:

"The family is purified, the mother becomes blessed,
The whole earth is rendered pure by a devotee
Whose heart and mind are absorbed in the boundless
Ocean of Existence-Knowledge-Bliss."

Sri Krishna says to his disciple Uddhava in the Bhagavatam: "He who loves Me is made pure; his heart melts in joy... melts in love... such a devotee is a purifying influence upon the whole universe." The devotee's capacity to purify others is powerfully expressed by Sri Krishna in Bhagavatam when He says that he himself always follows His devotees wherever they go, so that He may purify Himself by the dust of their holy feet (11. 14. 16).

The greater the devotee, the wider his sphere of spiritual influences. Even if he lives the life of the recluse in a cave, the vibrations from his devotion will spread over the whole world, and find an echo in all pure hearts ready to receive them.

Sanctifying Holy Places: A devotee immersed in Supreme Devotion has attained the highest state of sanctity and holiness. Sacred places can no longer sanctify him; rather they crave for his footprints (N.B.S., 69). In fact, the places visited by such sages and saints become pilgrimage centres, and those that are already holy places become more glorified. It is not the waters or idols of those places that really make them holy; but the saints who purify by virtue of God residing in their hearts. (Bhagavatam, 10. 48. 31)

Rendering Scriptures Authoritative: A Supreme Devotee has risen above the scriptures. Their teachings have borne the fruit of Divine Love in his personality. Therefore it is the Devotee who now gives authority to the scriptures (N.B.S., 69). He is a living commentary to the Gita, the Upanishads, the Bible and other sacred texts of the world. Without the radiant example of saints and sages, the scriptures would be a mere collection of empty words.

Full of God: A devotee supremely blessed is filled with the Spirit of God (N.B.S., 70). He has no room for his personal ego. His will has merged in the Will of God. Thanks to his self-surrender he has no independent existence for all practical purposes. He has become a living temple of God. Whatever he does becomes glorified, wherever he goes becomes sanctified, whatever he speaks becomes the contents of the sacred scriptures. God lives and moves through him.

Cosmic Celebration: God-realization by a devotee is the greatest event in this world. Divine Enlightenment heralds the forces of light that destroys the forces of darkness.

Mother Earth becomes afflicted when there is a rise in unrighteousness and decline in righteousness, when Divine forces are oppressed by demonic forces, and when the path of Spiritual Enlightenment is lost in the overgrowing forest of materialism. Afflicted with this burden she prays to the Lord for her rescue. God then responds by incarnating on the earth or manifesting through Sages and Saints. No wonder that there is a cosmic celebration to commemorate this greatest event (N.B.S., 71). Mother Earth is blessed with her saviour. The spirits of the devotee's forefathers rejoice due to the spiritual vibrations emanating from the devotee. The gods dance with joy because they are assured that the Divine force has re-emerged.

11

Qualifications and Qualities of a Devotee

Who feels the Need for God?

Not everyone feels the need for God. Many are satisfied with what the world of mind and senses can offer. However, there comes a time when, through the process of growth and evolution, or through frustration in life, everyone feels the need for God and becomes a devotee.

Kinds of Devotees: What kinds of people seek God? In the Bhagavad Gita, Sri Krishna states that four kinds of people worship God: the distressed, the seeker of wealth, the inquisitive and the wise (7. 16).

The Distressed:. The distressed are those who suffer from serious disasters, such as incurable diseases, natural calamities or attacks by a wayside robber, enemy or wild animal, or any other adversity. Being in a helpless position, he craves for the Grace of God. He calls, "O God! This is a lower form of devotion to God. A classical example of the distressed devotee was Draupadi, the wife of the Pandavas in the story of Mahabharatam:

Their evil-minded cousins, the Kauravas, tricked them into a fraudulent game of dice. Yudhisthira, the eldest Pandava, lost all his possessions, even his wife Draupadi. Seeing that the Pandavas were utterly helpless and had become their slaves, Duryodhana, the eldest Kaurava, wanted to hurt them further by insulting their wife.

Draupadi was dragged into the Royal Court where Duhshashana, one of Duryodhana's younger brothers, began stripping her of her sari (6 yards long cloth). In a state of extreme distress, she turned to the Pandavas, but realizing they could not come to her rescue, she turned her mind toward Krishna, the God-incarnate, and threw her arms up in surrender to the Lord. Suddenly, a miracle happened: as Duhshashana went on pulling Draupadi's sari, more and more cloth continued to unfold such that she was always seen in an ever increasing sari cloth. Thus Divine Grace saved her honour.

Similarly, in an Indian fable, there was an elephant called Gajendra. He called on Lord Narayana, as a crocodile was dragging him into the water, and he received the Lord's help.

The Seeker of Wealth: This person longs for earthly possessions, money and other forms of wealth in order to lead a happy life. He propitiates God to get His Grace to amass wealth. He may also crave for position, children, name and fame. This is also a lower form of Bhakti. However, as a devotee advances, his devotion is no longer directed toward some limited object, but toward God-realization.

Dhruva, a young prince, was deeply hurt by his stepmother's insult and practiced intensive austerity and devotion to God in order to become the successor to the throne. (For details, refer to the Bhagavatam.) When he had the Divine Vision he did not seek kingdom, but God's Devotion only.

The Inquisitive or Inquirer: He is a seeker after knowledge. He is dissatisfied with the world. There is a void in his life. He feels that the worldly pleasure is not the highest form of happiness. He prays for God's Grace in order to get wisdom. He eventually is able to discriminate between the relative and the absolute and seeks the superior knowledge of the Ultimate Reality.

In the Bhagavatam, Uddhava, a disciple of Sri Krishna, was dissatisfied with the world and sought wisdom from the Lord. All spiritual aspirants crave for God's Grace for attaining enlightenment. This is a higher form of devotion and this eventually leads to pure love for God.

The Wise: The wise man or the sage has realized the Self and is contented in the Self. He has no desires. He is freed from desires and greed. He finds his Self as the Cosmic Self or God. He is firmly established in Self and he has attained Oneness with the Divine. Nothing remains to be attained by him. Such a wise one worships just for the joy of worshipping. He is the finest of all. He is dear to the Lord and the Lord is dear to him. He is called Mahatma. Such great souls are rare, and they become torchbearers and examples to all other aspirants. Because of these great Mahatmas, the Spiritual Light always shines.

The highest devotees are those who have one-pointed love for God, and Love for Love's sake only (N.B.S., 67). In the Bhagavatam, Sri Krishna says to His disciple Uddhava:

> "To one, who finds delight in Me alone, who is self-controlled and even-minded, having no longing in his heart but for Me, the whole universe is full of bliss. Neither the position of Brahma nor that of Indra, neither the domination over the whole world, nor occult power, nor even salvation is desired by the devotee who has surrendered himself unto Me and who finds bliss in Me."

This indeed is the transcendental pure love. The great Commandment in the Bible describes it as:

"Thou shall love the Lord, the God with all thy heart, And with all thy soul, and with all thy mind."

The highest devotees always direct their thoughts toward God and seek Him with intense longing in their hearts. They dwell in God. They not only see God in their own hearts and realize their oneness with Him, but also see the same Lord in the hearts of all, and serve God in all beings, knowing His oneness with all.

Sri Krishna speaks of all four kinds of devotees as "noble" indeed. No matter with what motive a person begins to worship God and devote himself to God, once he begins to taste the joy in His worship, all other desires leave him, as illustrated by the case of Dhruva, mentioned earlier.

Jivakotis and Isvarakotis

Sri Ramakrishna classifies devotees into two categories: Jivakotis (ordinary persons) and Isvarakotis (Divine Messengers). By virtue of spiritual practice and devotion to God an ordinary person becomes a Jivakoti. The Jivakoti's devotion is formal; conforming to scriptural laws, such as repeating God's name a specified number of times. This kind of devotion leads to the knowledge of God and ultimately to Samadhi (Absorption). The Jivakoti does not return from Samadhi to the relative plane.

An Incarnation of God or one born with some characteristics of an Incarnation is called an Isvarakoti. The Isvarakoti follows the process of negation and affirmation. First he negates the world, realizing that it is not Brahman, but then he affirms the same world, seeing it as the manifestation of Brahman. Sukadeva, the narrator of the Bhagavatam and an ideal monk, was an Isvarakoti.

An Isvarakoti can return to the plane of relative consciousness after attaining Samadhi. An Incarnation is holding in His hand the key to others' liberation. And so, for the welfare of humanity the Incarnation returns from Samadhi to the conscious plane.

The Isvarakoti is like the king's son. He has the key to all the rooms of the seven-story palace; he can climb to all the seven floors and come down at will. A Jivakoti is like a petty officer. He can enter some of the rooms of the palace; that is his limit. The Isvarakotis can liberate themselves whenever they want to; but the Jivakotis cannot do so. (The Gospel of Sri Ramakrishna, 1986, pp. 237, 708, 749, 777)

The Qualifications of a Devotee

The qualifications to be possessed by an aspirant are described below:

1. Aspiration. Since God dwells in all beings, how can one identify a devotee? A devotee is a seeker of God. The chief qualification for an aspirant is to feel the need for God and to devote himself entirely to God. If a person is not interested in realizing God, it

does no good to lecture to him a hundred times. "You cannot," as Sri Ramakrishna says, "make any dent with a nail on a rock, however, you hammer on it. The nail will break. Similarly, there is no use in trying to teach a person about God if he is steeped in worldliness." A prospective aspirant should have an interest in spirituality and an aspiration to realize God.

In this respect, the Path of Devotion is different from other Paths. For instance, to follow the Path of Knowledge, an aspirant should possess discrimination, dispassion and several other virtues. A devotee, on the other hand, does not need to suppress any of his emotions, but to intensify them and direct them to God. Everyone, even a worst sinner, can worship God. If the most sinful one worships God with undivided attention, he too should be deemed righteous, for he has made a holy resolution to give up the evil ways of his life. Ratnakar, a thug, became the Sage Valmiki by his holy resolution. Sin vanishes when thoughts of God arise in the mind. In the Bhagavad Gita, Sri Krishna affirms:

"Even if the most sinful worships Me, with devotion to none else, He too should indeed be regarded as righteous, for he has rightly resolved Soon he becomes righteous and attains to eternal peace…" (9. 30, 31)

Even illiteracy is no bar. Saints like Kabir, Nanak, Tukaram, Sri Ramakrishna among others, had no formal education. Sage Sandilya says, "Everyone from the highest to the lowest-born is eligible to follow the Path of Devotion." Although no specific qualifications are needed, there is no doubt, however, that the seeker on the Path of Devotion will receive a greater benefit if he has equipped himself with virtues such as contentment, truthfulness, purity, kindness, charity, spirit of service, and the like.

1. Faith. A devotee must have unwavering and unshakable faith in God. He derives all his strength and courage from this faith. This faith removes all anxieties and uneasiness in his mind. What does this mean?

Faith means complete and full trust in the protecting power of God. The devotee has faith that everything happens for his own good and comes through the Grace of God. Both happiness and sorrow are forms of God's Grace. This kind of faith and devotion is not blind; it is built on the reality of God.

Faith is the root of religion. One cannot prove God if one has no faith in God. Faith without understanding is blind faith. It should be turned into rational faith. You will only experience that which you strongly believe. If you have no faith in God, or in your Divine nature, you never reach perfection. "One who thinks that Brahman does not exist, himself becomes non-existent," says Taittiriya Upanishad. Faith is the fundamental necessity for spiritual practice. Lord Krishna, in Bhagavad Gita, says:

"Those who, fixing their minds on Me, worship Me, ever steadfast
And endowed with supreme faith, are the best Yogis." (12. 2)

Therefore, faith in the Grace of God is essential for following the Path of Devotion. This faith is a living force in the true devotee.

1. *Nonattachment to Body:* Worshipping a manifested God – God with a form and name – is easier. We are embodied souls. Each one of us is a living being with a body and name. So it is natural and easier to fix our mind on a Personal God like Rama, Krishna, Buddha or Jesus. But worshipping the unmanifested Brahman is, however, very difficult, because Brahman has no particular form or name. It is pure Consciousness or Awareness. It is an abstract, invisible Spirit. This is why Lord Krishna says in Bhagavad Gita:
   "Greater is their trouble whose minds are set on the Unmanifested;
   For the goal – the Unmanifested – is very difficult. For the embodied to reach." (12.5)

Ultimately, the worshipper of the manifested (Personal) God and the worshipper of the unmanifested Brahman reach the same goal. However, the latter path is very hard and arduous, because the aspirant has to give up attachment to the physical body from the beginning of his or her spiritual practice.

The Qualities of a True Devotee

Knowledge of a true devotee's qualities is of great help to other spiritual aspirants. Aspirants on the Path of Devotion can experience greater benefits by cultivating the following qualities:

1. *Loving Heart:* The first important quality of a true devotee is that he has a very soft loving heart. There is no pride, hatred, lust, anger, egoism, greed and arrogance in the ideal devotee of God. In order to reach perfection, an aspirant must aim at getting rid of all these defects. He ought to be filled with joy, bliss, happiness, cheerfulness and peace of mind.

2. *Compassion and non-injury:* A devotee, who is established in God, bears no ill will toward anyone; he looks upon all with love and great compassion; he is free from attachments to worldly objects and sensual pleasures; he has no sense of ego; he is balanced in pleasure and pain; and he forgives those who hurt him. In the Bhagavad Gita, Lord Krishna says:

"He who hates no creature, who is friendly and compassionate to all,
Who is free from attachment and egoism, balanced in
Pleasure and pain, and forgiving... is dear to Me." (12. 13, 14)
"He by whom the world is not agitated
And who cannot be agitated by the world,
And who is freed from joy, envy, fear and anxiety,
He is dear to Me." (12. 15)

A true devotee never injures any creature in thought, word or deed; no creature is afraid of him. Unafraid of the world, he feels that the world is his own Self. He never hurts others and is not hurt by the words or actions of others.

3. *Contentment:* The true devotee is ever content, knowing that all that happens to him are the effects of his own past actions. He fixes his mind and intellect on the Supreme Being. He has a clear conviction regarding the essential nature of Self. He has controlled his mind and senses and so has no cravings. This is what we learn from the following verse of the BhagavadGita:

"Ever content, steady in meditation, possessed of firm conviction, Self-controlled, with the mind and intellect dedicated to Me, He, my devotee, is dear to Me." (12. 14)

4. *Free from Cares and Worries:* Cares and worries do not come near the devotee of God. Just as a little baby sleeping in the protecting arms of its mother is not afraid of anything. Similarly, the devotee of God feels the protection of the Lord at all times, even in the most frightful situations. He does his duty and leaves all the worrying to God.

He is free from dependence. He is indifferent to the body, the senses, the objects of senses and their mutual connections. He has external and internal purity. He renounces all actions meant for securing the objects of enjoyment; he has abandoned all egoistic initiative in all actions; and he has merged his will in the Cosmic Will. Such a devotee "is dear to Me," says Lord Krishna, in the Bhagavad Gita:

"He who is free from wants, pure, expert, unconcerned, And free from pain, renouncing all undertakings or initiatives, Who is (thus) devoted to Me, is dear to Me." (12. 16)

- *Treats all Alike:* The devotee treats all alike. He does not see some as friends and others as enemies; all are equal in his vision. He does not see any difference between a Christian and a Hindu, an American and an African, a rich man and a poor man. His love extends to all alike. How is this possible? This is because the devotee sees all beings as the children of God. He sees his oneness with all beings.

- **Joy:** In expressing the joy of his heart, none excels the devotee of God. The Bhagavatam describes a God-intoxicated devotee as, "Sometimes he weeps in the loving remembrance of the Lord; sometimes he rejoices and laughs; sometimes he dances in indescribable Divine ecstasy; sometimes he sings melodiously the glory of the Lord; sometimes he enacts the noble deeds of the Lord's Incarnation; and sometimes he sits quietly and enjoys the highest bliss of the Self."

- **Balanced Mind:** A true devotee is neither elated by praise nor pained by censure. He keeps a balanced state of mind. He does not rejoice when he gets desirable objects, nor hates when he gets undesirable things. He does not desire the unattained. He is not swayed by the blind forces of attraction and repulsion. His mind is serene and calm, as he has controlled his thoughts. He has renounced both good and bad actions. Lord Krishna says:

"He who neither rejoices, nor hates, nor grieves, nor desires, Renouncing good and evil, and who is full of devotion, is Dear to Me." (*Bhagavad Gita*, 12.17)

- **Sincerity in Seeking:** The true devotee is sincere to the core. Sincerity in seeking comes only when you know that you have a limited capacity. It is only when you say, "I can't do it anymore, please help me," that the help comes. You then link yourself with an unlimited capacity. Sometimes God acts like a hard-hearted person. He waits until you really give up and renounce the last bit of your egoism. Until then, your prayers are not really sincere. They may be beautiful. You may sing nice songs. Your chants may be wonderful. But when you cry, you do not worry about your voice, about the tune, and about the pitch. You simply shout, "Oh my God, please have mercy on me!" And that is what you call sincere prayer. That prayer is heard immediately.

You cannot do everything using your own capacity. It is important to realize that there is a higher power, a grace, to help you; but you have to sincerely ask for it. That is why the Bible says, "Ask and it shall be given." Unless you ask, you will not receive it. Just by asking sincerely, you are opening yourself to the higher power.

God's Grace is everywhere. He is not selective about who he blesses. He and His creation, which we call Nature, are always ready to give you. You need not go and praise the sun to get light. Just open the window and the sunlight comes in. As long as you do not put up a barrier, you get light. In the same way, God is ready to help you. Do not allow your pride to get between you and God's Grace.

By ourselves, we can achieve very little, compared to what we can achieve with the help of God. We do not lose anything by giving ourselves to God, but have everything to gain. The great South Indian saint Manickavasagar said, "I gave myself to You, Lord, and in return You gave Yourself to me. Who is gaining? With me You are not gaining anything – just a useless thing, incapable of doing anything. I will be a burden to You, one more stomach to be fed.

But with You I can do many things. If I have You, I have gained everything in the world. So You see, though You call it a fair business deal, You are really losing."

Once in Southern India there lived a very orthodox Brahmin. Every day he spent several hours in his morning worship. He had a beautiful altar, and he used to perform a big Puja there. But at the same time he had a little weakness in his mind. He was very fond of a certain kind of sweet pudding. Unfortunately, his financial situation was such that he could not afford to have the pudding every day. But he was so keen on having it that he thought of a little trick to get some every day.

Normally, during the Puja, when the food is offered to God, a curtain is placed before the altar to allow Him to enjoy it in private. After a while, the pujari will go in and bring out the blessed food to share with everyone. So he asked his wife to make a little pudding to offer to God at the morning Puja. But when he went behind the curtain, he would eat the food himself, and then come out and tell the family that God accepted is worship and ate the whole pudding. And everybody in the family was so happy because they thought that every day God came and ate the offering.

So it went on this way for many months. One day it so happened that he had to go to a distant village for an important occasion. And as an orthodox man, he wanted the daily service at his home to continue. So he called his son, a little boy who used to help him with the Puja, and said, "Son, you know how I have been doing the Puja. I won't be able to do it today, and I would like you to do it."

And the son replied, "Okay, Dad. I am happy to do it."

So the father left, and the son began the Puja. As usual, the mother brought a little pudding and gave it to the son to offer to God. So at the proper time during the Puja, the son placed the pudding in front of God, closed the screen, and went out to wait while God ate the pudding. After a little while, he went to remove the bowl, and to his horror, he saw that the pudding was still there. So he thought, "I must have done something terribly wrong. God is not happy with my Puja, so he refused to eat the pudding today. I thought I was doing the Puja exactly like my dad." And then he cried out to God, "Please, God, if my father finds out that I made you to starve, he will be so angry with me. Please eat the pudding."

So he placed the bowl again, closed the screen, and went out to wait. When he returned, the bowl was still full. He fell in front of the altar and started crying, "God, if you don't eat the pudding today, and if my dad comes and finds out, he's going to kill me. Please eat it."

He checked again, and the pudding was still there. This time, he started crying so much that he fainted. After half an hour when he somehow got up, he saw the bowl was empty. He started jumping with joy. "Oh, God, I'm so happy that you heard my prayer. Now I can tell my dad that you accepted my service."

He was eagerly awaiting his father's return. His father came back very late that evening, and as soon as he saw him walking to the house, the son ran to meet him and said, "Dad. God accepted my service, too. He ate the food just as He does for you every day. First, He refused, but then I cried and cried and cried and made Him eat."

Now, the father thought that, somehow, the boy always knew what he was doing and was simply waiting for an opportunity to do the same thing. So he questioned him, "Are you sure that God ate the pudding?"

"Yes, Dad, there's no doubt about it."
"All right. In that case, you do the service tomorrow, also. I want to see it."
"Okay, Dad. If God ate the pudding today, why shouldn't He do it tomorrow, too?"

So the next day he did the service, but God did not eat the pudding. The boy started crying again. He was almost ready to smash his head on the floor, and at that point a voice came, "Son, don't cry. I would accept it, but not in front of your father. He didn't deserve to see me eat it. That's why I didn't eat it right away."

The father also heard the voice. He literally fell at the feet of the child and said, "You are my Guru. You've opened my eyes. I wish I had faith and devotion like you. I'm a scholar; I did the Puja correctly according to the scriptures, but I didn't have that kind of innocent faith that you have."

Even if you know the entire scriptures by heart, you will not gain God's Grace. God doesn't care about how much you know; He wants your pure heart. A sincere prayer from a pure, faithful heart will never go unanswered.

- *Enjoy the Company of Devotees:* Devotees always like to talk about God to other devotees. Sri Ramakrishna was used to becoming absorbed in Samadhi many times during the day and night, while conversing with the devotees. Though he lived in a constant vision of God, the Mother, he would long for the companionship of the devotees. When they talk about God, their voices choke, tears flow from their eyes, and their hairs stand erect in ecstasy (N.B.S., 68). Sri Chaitanya describes this ecstasy in a prayer:

"Ah, how I long for the day when in chanting Thy name,
The tears will spill down from my eyes,
And my throat will refuse to utter its prayers,
Choking and stammering with ecstasy,
When all the hairs of my body will stand erect with joy!"

12

Guru's Role

Who is a Guru?

The spiritual practice of Yoga, like a craft, requires the guidance of a Guru or teacher. The word *"guru"* is made up of two syllables: *"gu"* and *"ru." "Gu"* means the darkness of ignorance, and *"ru"* means one who dispels. So the one who dispels your ignorance is the Guru or spiritual teacher. He is one who helps you in realizing your own Spirit or *Atman* by removing the ignorance that veils It.

The Need for a Guru

Right from birth we learn through the help of somebody. For a baby the mother is the first Guru. It is through her the baby knows its father. The Hindu scriptures say that everyone has four Gurus: *mata, pita, Guru, Deva*—mother, father, the spiritual guide and God. First, mother shows you the father, then the father takes you to the Guru, and finally the Guru takes you to God.

Even for acquiring worldly knowledge, we take the help of many teachers. The need for a Guru is much more in the spiritual life, because the spiritual life is much more subtle and deeper.

One may read a lot of books on Yoga and spiritual life, but books can never take the place of a Guru. If we could know everything through books, there would be only publishing houses and no universities and ashrams. Books have information, and you can read them, but they cannot teach you. When you read a book, you may understand or misunderstand what you read. This depends upon your background, level of knowledge and comprehension. But a Guru will not allow you to misunderstand. The moment he finds you have not understood something correctly, he will correct you and also clarify your doubts. A book cannot do that. That is why you need a person who has gone through the path and realized the goal to guide you in treading the spiritual path. Therefore a spiritual teacher is essential. Yoga is not a theory but a practical method. You can learn its subtle aspects only through a spiritual guide.

The Guru guides the seeker on the Path of Devotion. The spring of devotion passes from the Guru to the disciple. The ultimate Truth can be revealed only by somebody who has experienced it. No candle can light itself; another lit candle must come and touch it. That is the way Truth passes from the Guru to the disciple. The Guru shows the path, the pitfalls on the path, and how to overcome them. It is only the Guru who can find out your defects. The nature of egoism is such that you will not be able to find out your own defects. Just as you cannot see your back, so also you cannot see your own errors. You must live under a Guru for the eradication of your bad qualities and defects.

Under the guidance of a Guru you are safe from being led astray. Association with the Guru is an armour and fortress to guard you against all temptations of evil forces of the material world.

Cases of those who had attained perfection without study under any Guru should not be cited as a justification against the need for a Guru; for such great persons are exceptions in the spiritual field. They had taken birth after having reached greater heights in spiritual evolution in the previous birth.

Spiritual development needs an impulse to action. The quickening impulse must come from another soul, the Guru. The soul from which it comes must possess the power of transmitting it to another; and the disciple to whom it is transmitted must be fit to receive it. Then alone, the spiritual growth will come.

How to Recognize the Guru?

In these days there are many Yogis who claim to be perfect Masters, and it is not easy to identify a true Guru from the impostors. In the *Kularnava Tantra*[1] Lord Shiva says,

> "Many are the gurus who are like lamps in homes,
> But rare is the Guru who illumines all like the sun.
> Many are the gurus, who are learned in the *Vedas* and the scriptures,
> But rare is the Guru who has attained the Supreme Truth."

If you look for a medical doctor, lawyer, psychiatrist or any other professional, then you can scan the yellow pages in a telephone directory. But you cannot do so to identify a true Guru. You should prepare yourself by developing mental capacities. When you start discriminating between right and wrong, and between real and unreal, you will be able to know the true teacher. The *Upanishads*, the ancient Yogic scriptures, state that when the disciple is ready, the Guru appears. This means that when you have longing for Divine Love, you will receive Divine help sooner or later.

Mysterious Divine Help

For a sincere aspirant help comes in a mysterious manner. Ekanath (sixteenth century)[2] heard a voice from the sky. It said, "See Janardan Pant at Devagiri. He will put you in the proper path and guide you." He acted accordingly and found his Guru. Tukaram (seventeenth century)[3] received his *mantra* "Rama Krishna Hari" in his dream. He repeated this *mantra* and had the *darsana* (vision) of Lord Krishna. Lord Krishna directed Namadev (thirteenth century[4] to get higher initiation from a monk at Mallikarjuna. Madhura Kavi saw a light in the firmament for three days consecutively. It guided him and took him to his Guru Nammalvar who was sitting in his *Samadhi* underneath a tamarind tree near Tirunelveli in Southern India. Tulasidas (sixteenth century)[5] received instruction from an invisible being to see Hanuman and, through Him, to get the Vision of Sri Rama.

The sun requires no torch to make it visible; we do not light a candle to see the sun; when the sun rises we instinctively become aware of its rising. Similarly when a Guru comes to help us, the soul instinctively knows that it has found the Truth. In the words of Swami Vivekananda, "Truth stands on its evidence; it does not require any other testimony to attest it; It is self-effulgent." Such are the very great teachers; we can get help from the lesser teachers also. How are we to know what he is?

The Qualities of a True Guru

How can you recognize a true Guru? What are his qualities? In the *Bhagavad Gita*, Arjuna (the disciple) asks Lord Krishna (the Guru) the very same questions, though using a slightly different word instead of "Guru," he asks:

> "O Krishna, what are the characteristics of the sage of steady wisdom,
> Who is established in the Super-conscious state? How does he sit.
> How does he speak? How does he walk?" (2. 54)

A Guru is a sage of steady wisdom (*stithapragnyam*), one who has attained Self-realization or enlightenment. However, it is very difficult to recognize a Guru because in outward appearance, he looks like any other person. It is his inner qualities that reveal his true nature.

Lord Krishna describes his qualities in the verses 55 to 57 in chapter two of the *Bhagavad Gita*. The Self-realized sage of steady wisdom is one who has renounced all the desires and is satisfied in the truth of Self. He is not perturbed by adversity. He does not crave for pleasure. He is free from attachment, fear and anger. He is poised in wisdom. He is neither delighted at receiving good nor hates pain. He is, like the ocean, totally contented. He has no wants. As he realized that the ever-changing external world is unreal and that the eternal *Atman* alone is real, the objects of the world have no value to him. The sage has withdrawn his senses from sense-objects. The *Guru Gita*[6] says that a fully enlightened being is *brahmanandam,* the bliss of the Absolute;

paramanandam, the giver of happiness at all times; and *kevalam jnanamurtim,* the embodiment of knowledge, the embodiment of the Truth.

How does the sage of steady wisdom conduct himself? What is his behaviour? The self-controlled sage, though moving among the sense-objects, is free from attraction and repulsion. His mind remains calm and balanced. He has no sorrow and anxiety. His peaceful mind is full of bliss. As his senses are under control, his wisdom is steady. What is real to the worldly-minded is illusion to the sage. He lives in the Self, unaware of the world. *(Bhagavad Gita,* 2. 64, 65; 2. 68, 69)

The ocean never wants anything. It never sends invitations to the rivers. The ocean is just there. The rivers fall in love with the ocean and run headlong toward it and enter into it from all sides. But the ocean does not swell up. In the same way, the steady minded sage remains unmoved even when every thing comes to him. *(Bhagavad Gita,* 2. 70)

How does a sage of steady wisdom look at things? He looks with equal vision at a *brahmin* endowed with knowledge and humility, a cow, an elephant, a dog and an outcast *(Gita,* 5. 18). He does not make distinctions like ordinary people who differentiate others in terms of language, skin, colour, caste and country. Like Nature or God, a steady minded sage has equal vision. He has a balanced mind toward well-wishers, friends, enemies, the indifferent, the impartial, the hateful, relatives, saints, as well as sinners *(Gita,* 6. 9). To him a clod of earth, a stone, and a piece of gold are the same *(Gita,* 6. 8).

A realized sage hates none, is friendly and compassionate to all, free from the feeling of "I and mine" and is balanced in pairs of opposites: pleasure and pain, cold and heat, honour and dishonour *(Gita,* 6. 7; 12. 13 and 12. 18). He is ever contented and self-controlled. He is free from joy, envy, fear and anxiety. He neither becomes excited nor depressed *(Gita,* 12. 15).

He is *Satchidananda* (Existence-Knowledge-Bliss) at all times. Nothing will shake him. He neither rejoices, nor hates, nor desires, as he has renounced both good and evil (*Gita,* 12. 17). A realized Guru possesses all the above qualities.

A true teacher is pure or sinless. A teacher of secular subjects like physical sciences may be of any character; all he needs is intellectual power. But for teaching spiritual sciences, the teacher should be pure. "Blessed are the pure in heart, for they shall see God." (New Testament Matt chapter 5 v.8) A vision of God never comes until the soul is pure. Therefore, in the Guru we must look for purity.

The Qualifications of a Guru

The traditional qualifications of a Guru have been laid down in the *Mundaka* and *Brahadaranyaka Upanishads*. Shankaracharya in his *Vivekacudamani* (Crest-jewel of Discrimination) has summarized these qualities:

1. The Guru must be versed in the scriptures. But he need not be a scholar. He must know the spirit of the scriptures. Sri Ramakrishna used to tell a story: a few men once went into a mango orchard. Soon after, most of them were busy counting the leaves, examining the colours of the leaves, the size of the twigs, the number of branches and so forth. They began to note down everything, and started a wonderful discussion on each of their observations. But one of them, more sensible than others, kept himself apart and busy eating the mangoes all the time. Who was the cleverest of them all? This type of activity, counting of leaves, twigs and branches, and note taking, has its own value in its proper place, but not in the spiritual field. Human beings never become spiritual through such work; no one has ever seen a strong spiritual person among these "leaf-counters." Spirituality is the highest aim for us; it does not require knowledge of the entire scriptures word by word. One only needs to feel the spirit of the scriptures. One who deals too much in words, loses the spirit. The various methods of speaking a beautiful language and explaining the diction of the scriptures are only for the

enjoyment of the learned; these do not take one to perfection. No great teacher goes this way.

2. The Guru must be established in *Brahman*, ever living in God. He must have attained Self-realization, for if he has not experienced the Truth, how can he carry us to the shore beyond ignorance? The subject taught in the spiritual pursuit is God Himself, the Ultimate Truth. The Guru must, therefore, be able to show us how to manifest the Divine Being within.

3. The Guru should be untouched by desire, unmotivated by any selfish purpose. He should have no hankering for lust, name, fame and honour and money. He should be totally unattached.

4. The Guru should be stainless, guileless. The genuine Guru never thinks of himself as a Guru. He puts on no airs. He should have no sin, no conceit, no deceit, and no deviousness in him.

5. The spiritual Guru ought to be a renunciate. If he owns private property somewhere, has financial investments or is worrying about spouse or family, how effective can he be as a spiritual teacher?

6. What is the motive behind the Guru? Does he teach with any ulterior motive of name, fame or gain; or does he do it for pure love for his disciples? Spiritual forces can only be conveyed through the medium of love; no other medium can convey them. Any other motive such as name or fame would destroy the conveying medium.

Guru's Attitude toward his Disciples

What is the Guru's attitude toward his disciples? Does he advertise in yellow pages and invite people to become his disciples? No genuine Guru is interested in creating disciples. He will not go out preaching; he is not a missionary. In fact no genuine sage will ever declare himself a Guru. It is the disciples who recognize him as the Guru. They make him a Guru. They seek his guidance and call him a teacher. Otherwise he is just there. "Ask it shall be given," says the *Bible*. Knowing the fitness of the student and his eagerness to receive, the Guru gives to him. Otherwise he remains as he is. A true Guru does not need advertisements. Have you ever seen a flower sending out a circular to all the bees saying, "I have honey, come to me?" No. When the flower is in bloom, the bees will come. So when a person really becomes tired of the worldly pursuits and looks for eternal peace and joy, he seeks the guidance of the Guru.

The Guru Within

One should not take the physical body or the intelligence of a teacher as the Guru. The Self is the real Guru, because he has realized the Self. His intellect gets a better light from the Self. The scriptures say, "Guru is Lord, Guru is Divine, Guru is your relations, Guru is your body, Guru is your soul, Guru is your Self. There is nothing but the Guru." This means ultimately everything is that Self.

The real Guru is the Spirit within you, the Awareness. It is your own conscience. It is a part of the Cosmic Consciousness. It is the God within you that is always watching you. It can guide you, tell you whether you are doing right or wrong. It is the Guru within you.

But sometimes you may be weak and not listen to that conscience. So you have an outside Guru who has realized the inner Truth and who follows his conscience every time. That Guru helps you to know what is right and what is wrong. Sometimes you have doubts as to which inner voice to listen to. Your ego says, "Go ahead and do it." The conscience says, "No, that is not the right thing to do." But you do not know which is the ego and which is the conscience. The Guru helps you to get that doubt cleared. Even while helping you, he will gradually teach you how to recognize and follow the inner Guru. He will never make you dependent on him. A Guru is there to liberate you. Spiritual growth means freedom from bondage. No Guru will ever ask you to always be in his physical presence. There are many disciples who rarely even see the Guru physically, but still understand his teachings.

That is possible when you look within, meditate, and see him inside. Then you get all the answers from him.

The teaching is the real teacher. The Truth is eternal. It is just transmitted to you through a physical body. The help that can be given through a physical body is limited; but on the spiritual level, the Guru can help you much more. In your practices, in your prayer, you should learn to communicate with the spiritual part of the Guru. Always try to communicate at that higher level.

Subsidiary Gurus

When the mind is purified through moral life, prayer, meditation, etc., it comes in touch with the inner Light of the Supreme Spirit. The purified mind becomes a channel for the flow of Divine Knowledge. It receives spiritual guidance directly from the Teacher of teachers. When the mind learns to open itself to the inner Voice, it can receive instructions from many sources. The *Bhagavatam* speaks of a wandering ascetic who learnt from twenty-four natural objects. From mother earth he learnt the secret of patience, from air he learnt detachment (as air remains unaffected by pleasant or bad odours), from the sky he learnt freedom from limitations, and so on (2. 7. 9).

Brother Lawrence, the seventeenth-century French mystic, spent his life in the kitchen of a monastery. The sight of a leafless tree in mid-winter stirred in him the reflection that leaves would be renewed, and flowers would appear on its branches. This revealed to him the presence and power of God lying hidden in all creation. The spiritual awakening that he experienced sustained him throughout his life.

Guru's Role

In the spiritual field, the Guru's role is unique. You are essentially Divine by nature. You have forgotten your true Divine essence. So instead of realizing that, you get drowned in world-process more and more. Somebody must remind you of your real nature. It is the Guru who does this. He awakens you from your age-long sleep and shows you the way to the Divine. He removes your ignorance and enlightens you. He studies your temperament, eagerness, outlook and level of spiritual development, and prescribes appropriate path and methods. He assesses your progress, corrects your imperfections, and guides you through the practice. He is a channel for power that flows down the long line of Gurus dating back many thousands of years.

Sri Ramakrishna says, "One must have an awakening of the Spirit within in order to see the unchanging imperishable Reality." How is this first awakening brought about? An illumined Guru does this for you through a process of spiritual initiation. The Guru transmits the word of God called *mantra* to you. The power of God comes through the *mantra*, and through the *mantra* comes the awakening of the Spirit. The power of initiation manifests only in a pure soul who intensely yearns for Self-realization. It is not enough to get directions from the Guru; you must struggle incessantly; you must purify through cultivation of virtues and devotion to God; and you must yearn with all your heart to know the Truth.

The initiation is the first stage in guidance into the spiritual process. Initiation is said to be of three kinds: *shakti* (power), *sambhavi* (touch or look) and *mantra* (sacred word).

Shakti. In the *shakti* initiation, by his mere will the teacher wakes up the dormant spiritual knowledge in you, giving you full or partial realization right at that particular moment. He does not even have to touch you or be near you. Generally only an *Avatar* (Incarnation) has the power to give this type of initiation. In the life of Sri Ramakrishna this power was seen on various occasions.

Sambhavi. In this initiation, by a touch or look the Guru imparts spirituality. This, too, is unusual, and only great saints have this faculty.

Mantra. All other initiations are by *mantra*, in which the Guru utters a sacred word into your ear, thus planting in your mind, as it were, the seed that sprouts into spiritual unfoldment. (For further details, *see* chapter 14. Mantra) This is like putting a drop of curd into a pot of milk in order to ferment and convert the milk into curd quickly. Sometimes an aspirant may receive initiation from a great saint in his dream. A truly inspired dream-experience is as sound as a personal initiation.

The advanced stages include initiation into *Pre-sanyasa* (Pre-monkhood stage) and initiation into *Sanyasa* (Monkhood stage). These initiations are given to those who desire to renounce the worldly life and dedicate their entire life to spiritual practice and service.

The Guru will guide you in practicing advanced techniques of *Pranayama* and Meditation that ultimately lead you to the Super-conscious experiences. His illustrious life itself serves as a source of illumination, light, inspiration and message. The Guru removes the veil of ignorance that is covering your knowledge. He guides you until you begin to follow your conscience that is the inner Guru. Be worthy of the Guru's blessings and follow his directions sincerely in order to make steady progress in your inward journey. Gradually, through the Guru's blessings, the hidden faculty of intuition will awaken in you, and your purified intuition will act as your Guru.

Often the role of the Guru is misunderstood, and the disciples think that the teacher would carry them to the goal. If you see a signpost that says, "New York 200 miles," you would not prostrate before the signpost, deck it with garlands and pray, "Please take me to New York." The signpost simply points out the direction and the distance to that place. You must do the walking. The Guru is like that. He has gone the route. He knows the journey and is able to guide you on the trip. But you have to follow his teachings. The Guru teaches through his personal example. His every day conduct is a living ideal to you. Observe his living example and imbibe his virtues. The Guru teaches according to your temperament and evolution. It is meant for you only. Do not try to impose on others the teaching you have received. You can imbibe or draw from the Guru in proportion to the degree of your faith and attention.

Three Aids

There are three aids to realize the Truth or the Ultimate Reality: the scriptures, the Guru and the spiritual practice. The scriptures tell you that sugar is sweet. The Guru will show you that sugar. Your practice will give you the taste. The Guru will not put the sugar in your mouth and say, "It is sweet." You have to taste it yourself. Even if you open your mouth and he puts the sugar in it, if your tongue is totally coated, you cannot taste it. So you must be fit to experience the bliss of God.

Sometimes when the disciple is fit for tasting the spiritual experience, the Guru may transmit that experience to the disciple. Sri Ramakrishna did this to Swami Vivekananda (then called Narendra). Sri Ramakrishna touched his disciple and Narendra went into *Samadhi*. He tasted the experience of the state of Super-consciousness. To get such a little glimpse, the disciple needs a lot of sincerity, purity of heart and devotion. Of course, he has to practice intensively to gain that experience on a regular basis.

Guru's Tests

The Guru may test the students in various ways. Sometimes he may even tempt them. Some misunderstand him and lose their faith in him. In ancient days, the tests were very severe. Once a Guru named Goraknath asked some of his students to climb up a tall tree and throw themselves, head downward, on a very sharp trident (*Trisula*). Many faithless kept quiet. But one faithful student at once climbed up the tree with lightning speed and hurled himself downward. The invisible hands of Goraknath protected him. The student then had immediate Self-realization.

Avatars as Guru

Besides realized sages or enlightened seers, there are the *Avatars* (God coming in the form of a human), the Teachers of all teachers. They are much higher and. can transmit spiritual experience with a touch, with a wish, which makes even the most degraded sinners saints in a second. Sri Krishna, Buddha, Christ, Sri Ramakrishna and other Incarnations are these greatest manifestations of God. We cannot see God except through the Incarnations.

When an *Avatar* touches a person, his whole soul will change; he will be transformed into just what the *Avatar* is; his whole life will be spiritualized; and spiritual power will emanate from every part of his body. The *Avatar* can alter the destiny of people, and wipe out their karma. No ordinary teacher has such a power of transforming a person.

Jesus had the power to bring Divine Light to simple fishermen at his touch. He also had the power to transform impure souls whom people called sinners. When he told them, "Thy sins are forgiven; thy faith has made thee whole; go in peace," they at once felt freed from all impurities.

In modern times Sri Ramakrishna was an *Avatar*. He had the power to raise disciples to great heights of illumination by a mere touch or look or wish. As mentioned earlier when Sri Ramakrishna touched Narendra, he immediately got Super-conscious experience.

The great strength of an *Avatar* does not lie in His miracles or His healing power; even devils can do that; but it lies in His teachings. So in worshipping an *Avatar*, in praying to Him, we must always remember what we are seeking. "Not those foolish things of miracle-display," says Swami Vivekananda, "but the wonderful powers of the Spirit, which make man free, give him control over the whole of Nature, take off from him the badge of slavery, and reveal God unto him." (*Religion of Love*, 1970, p.47)

The Eternal Teacher

Real initiation takes place when God awakens your spiritual consciousness. The real Guru is the indwelling Supreme Spirit that is "the Goal, the Controller, the Lord, the Witness, the Abode, the Refuge, the Friend, the Origin and Dissolution of the Universe, its Substratum, the Repository of all Knowledge and the Eternal Seed," (*Gita*, 9. 18).

The disciples look upon the Guru as a visible manifestation of the Supreme Spirit, the Teacher of teachers, as a channel for the flow of Divine Grace. It is in this spirit that they serve him, obey him and worship him. The great sage Sankaracharya in his *Daksinamurthi Strotum* says:

> "I bow to the Divine Guru who, by the application of the
> Collyrium of Knowledge, opens the eyes of one bound by the
> Disease of ignorance. I bow to the Divine Guru who imparts to the
> Disciple the fire of Self-knowledge and burns his bonds of karma
> Accumulated through many births.
> "I offer my salutations to the beneficent Being who is incarnate in the
> Guru,
> The Light of whose absolute Existence shines forth in the world of
> appearance,
> Who instructs the disciple with the holy text, 'That thou art,' realizing
> whom
> The soul never returns to the ocean of birth and death." (3)

REFERENCES

1. This is a treatise about Guru, the disciple, the *mantra* and different forms of worship.
2. A great mystical saint of Maharashtra, India. His famous lyrics called *"abhangs"* expresses his intense devotion to God.
3. The most popular saint of Maharashtra, India. He was a poet and preacher. His *abhangs* reveal the greater inner struggles he had to pass through in order to attain full illumination
4. A great saint of Maharashtra, India. Through intense devotion and constant repetition of Lord Vittala he attained God-realization.
5. An outstanding poet-saint of Northern India. A great devotee of Lord Rama. His immortal *Ramacaritamanas* (story of Rama) is a household scripture in Northern India.
6. *Guru Gita* is an ancient Sanskrit text that describes the nature of the Guru, Guru-disciple relationship and meditation on the Guru.

13

Discipleship

A disciple is one who seeks spiritual guidance from a Guru. He is a seeker of Enlightenment. Discipleship is very difficult to achieve, because it needs wholehearted faith, devotion and intelligent discrimination. The poet Bholenath (nineteenth century)[1] sings,

> "Anyone can easily become a Guru. But what is very hard is to
> Become a disciple . . . No one is ready to receive the teachings;
> But everyone feels clever enough to give teachings."

So it is rare to find a true disciple. It is easy to become a Guru, but it is so difficult to become a disciple. Nature is constantly giving the teachings, yet no one goes to Nature to receive them.

Three categories of Disciples

In the ancient Tamil writings we come across a wonderful description of the different kinds of disciples.

The Top Level. The top-level disciples are those who behave like a swan or a cow. What is one great quality that we see in the swan? If you mix water and milk and place it in front of the swan, the swan will put its beak into it, stir the whole thing and be able to extract the milk and leave the water behind.

The cow has a different quality, but it is as good as that of the swan. She will go and graze in one place, eating as much as she can with full concentration. Then she will go back to the stable, lie down and regurgitate everything she swallowed. She will keep it in the mouth and chew it well. Anything that is raw and not digestible, not good for the system, she will spit out. Finally, she will swallow the good part, digest and assimilate it.

A top-level student will have these qualities. He will receive everything, but will have the capacity to separate what is conducive for his growth and what is not. What is useful, he will put to use. He will assimilate it. That kind of discriminating faculty is found in the top grade disciple.

The Middle Grade. The disciple of middle grade will be like a parrot or the earth. A parrot repeats whatever it hears. It does not expound on it or develop it; it simply repeats. What is the quality of the earth? Whatever you sow, it gives back. It does not give anything else. If you sow a mango seed, you get a mango tree. If you sow an apple seed, you get an apple tree. Similarly, a middle level student simply repeats whatever he learns; he does not develop it.

The Low Level. For the low-level disciple there are four examples. First is a pot with holes. Whatever you pour in it immediately flows out. It does not retain anything. The next example is a buffalo. If you take a buffalo to a beautiful clear pool of water, it will not stay out of the water to drink. Instead, it will walk in, lie down, and stir up the whole thing. It turns the clear water into muddy water, and then just lies there and sleeps. This type of student will make a mess of whatever he hears.

The third example is the goat. The goat never stays in one place to eat. It will go and bite a leaf here, run and bite a leaf somewhere else, run again and bite something else. There are students like that. They will stay in a spiritual center for one week, then go to another center, and then do something else. They constantly move around, wanting quick results; they do not stay in one place long enough to learn anything.

Finally, there is the strainer. You pour the tea, and the strainer keeps all the unwanted dusts, and lets the nice tea run through. All the good things are missed, and all the undesirable things are held tight.

Qualifications of a Disciple

Who can find a Guru? One who is fit to receive spiritual guidance and is ready for it will find a Guru. What are the qualifications of a fitting disciple? He should have sincerity and acknowledge that he knows nothing (in the spiritual sense). He should be an "empty" cup. He should not say, "I do know something; can you add a little more?" When Arjuna was standing on his chariot in the midst of the battlefield, he was overwhelmed and deluded by emotions. He did not want to do his duty and fight, but argued with the Lord Krishna with all kinds of philosophical arguments to support his position. But ultimately he realized his foolishness and said,

> "I am overwhelmed by emotions. My mind is confused as to my duty. I entreat You to say decisively what is good for me. I am Your disciple. I have taken refuge in You. Please instruct me." (*Gita*, 2. 7)

That is what you call total surrender. You accept your ignorance. Then you are totally free from egoism. You come with an empty vessel. As long as your ego is in the vessel, whatever the Guru puts in would be contaminated.

When will an aspirant be ready for initiation? Here is a story.

Once a seeker went to see Lord Buddha and asked for initiation. Buddha said, "You are not ready for it yet."

The seeker asked, "Why am I not ready?"

"Because you are still affected by everything."

"How can I get over being affected by everything?"

"Come tomorrow."

The seeker returned the next day, and Lord Buddha said to him, "Go to the graveyard and praise the dead. Use the most flattering terms you can think of. Praise them to the utmost."

The seeker said, "All right." He went to the graveyard and started praising every ancestor, every relative. "How wonderful you were! How beautiful! How great! How incredible! How superb!" When he returned, Lord Buddha said, "Now go and yell at them. Scream at the top of your lungs. Revile them for everything bad that they have done."

The seeker thought, "Well, I asked for it." He went back to the graveyard. He started yelling and shouting at them. He used every foul word he could think of.

Again, when he went back, Lord Buddha asked him, "What did they say to you when you praised them?"

The seeker said, "Nothing."

"What did they say to you when you yelled at them?"

"Nothing."

Lord Buddha said, "When you become nothing, you will be fit for initiation. Go. Become nothing."

There is another wonderful Zen story of a proud young aspirant who went to a Master for the first time. Upon seeing the student, the Master, without saying anything, invited him for a cup of tea. The lad happily started talking about his accomplishments to the Master, and as they were talking, the Master began to pour tea into the lad's cup. But instead of filling it and stopping, he kept on pouring even when the cup was flowing over. The aspirant stared in amazement as the tea was brimming over the top. Finally unable to contain himself, he blurted out, "Master, it's already full and the tea is flowing out on the ground; you cannot put any more." The Master stopped and looked up, and said, "Well, it is the same with you, my son. You are filled to the brim with your own ideas and self-importance. Whatever I say is going to overflow and can't go in, until you empty yourself first."

Unless we overcome the ego and the attachment to pride, we cannot fully understand what the Guru teaches, and more importantly the spirit of his teaching. So cutting through pride and self-importance is a major step in our attitude towards our Guru. Early in our association, the Guru may impart many ideas and precepts to us, but only after many hearings and subsequent meetings we begin to sniff at the scent of Truth. Often, at times the Guru relates many different thoughts. But beneath all those ideas, what he is really saying, remains constant. What he is saying is not translatable into words; rather it is an inner message, an inner communication that is perceived when we empty ourselves and humble ourselves before him; otherwise we will never hear what the Guru is saying.

A seeker of Enlightenment should feel and say, "I'm just empty, hollow. You are holy. Please pour that holiness into this hollowness. I am a 'holey' reed, please play your music through it." A flute is just a reed full of holes. It does not have anything inside and therefore whatever the flutist wants to play, he can. A disciple should be like that.

The Conditions for Fitness. Who is fit to receive the guidance and grace of the Guru? One who satisfies the following conditions:

1. The first condition is that he should become nothing. He should feel that he knows nothing. He should have no sense of individuality, and no sense of self-importance.
2. The second condition is purity. No impure soul can be religious. Purity in thought, word and action is absolutely necessary.
3. The third condition is a real thirst for spiritual knowledge. One, who wants, gets. Wanting spiritual knowledge is a very difficult thing, not as easy as we generally think. Spiritual knowledge does not consist in hearing talks or reading books; rather it is a continuous struggle, a grappling with our own nature, a continuous fight, till the victory is achieved.
4. The fourth condition is perseverance. The struggle for overcoming ego and purifying the mind is not a question of one or two days, of years, or of lives, but it may even be hundreds of life-times, and the disciple must be ready for that. The disciple, who sets out with such a spirit and determination and practices Yoga with perseverance, finds success.

Thus a seeker must become fit for discipleship through his own efforts. He must develop the capacity to understand the implications of even what the Guru would be saying in an indirect mystical way.

Once there was a seeker called Malik-I-Dinar. He wanted to find a great Guru who could give him the knowledge of the Truth. He set out on his journey. On the way he came across a dervish

(Islamic Mystic saint). He asked the dervish where he was going, and the dervish said, "I'm just walking. Where are you going?"

"Well," said Malik-I-Dinar, "I'm in search of a Guru who can give me true knowledge." The dervish listened to him. Then Malik asked, "Do you know anything of this knowledge?"

"I don't know anything about anything. I'm just walking. If you wish to walk with me, you may join me."

Malik said, "All right. I'll join you." So they walked on together.

In the course of their journey, they came across a tree. The dervish put his ear against the bark, and said, "The tree is telling me that there is a thorn inside it that we should remove, because it is painful."

Malik said, "I'm in search of true knowledge! What do I have to do with this tree and its thorn? Let's get going!"

The dervish said, "All right." They kept on walking.

Night fell, so they found a place to sleep. When they woke up in the morning, Malik was very hungry. He told the dervish, and the dervish said, "I smelled honey in that tree. Let's go back and see if there is any honey there." So they went back and found a crowd of travelers, all drinking honey from the tree.

Malik approached one of them and asked, "How did you know there was honey in this tree?"

"We didn't know," he said, "but we heard the dervish telling you that a thorn was hurting the tree, so after you both left, we removed the thorn, and all this honey gushed out."

"Oh," said Malik,

The dervish and Malik were hungry, so the travelers gave them some honey. They had their fill, and again set out on their journey.

They came across an anthill. The dervish knelt and put his ear close to it. He told Malik, "The ants are telling me that there is a rock obstructing them from building their home, and asking us to remove it."

Malik looked at the dervish and said, "You know, I'm in search of the highest knowledge. These ants and their rock have nothing to do with me! Let's not waste any more time here. Let's get going."

"Very well," the dervish said. So they left.

Once again they made camp for the night, and when they woke up in the morning, Malik said, "I have lost my precious knife! Have you seen it? I have to have it on my journey."

"You might have dropped it at the anthill," said the dervish, "Let's go back and find it." So they came back to the anthill. What did they find there? A crowd of travelers distributing gold coins.

Malik went up to one of them and asked, "Where did you find the gold coins?"

"In the anthill," he replied.

"What?" said Malik.

"Yes," he said, "We were walking by and we overheard the dervish telling you that a rock was obstructing the ants from building their home. Out of kindness we removed the rock, and under it we found a pot filled with these beautiful gold coins."

The whole thing amused Malik. The travelers were so kind that they gave some gold coins to Malik and the dervish.

Malik began to think: "Twice I missed out something of great value. But what do I have to do with the tree and its thorn and honey? What do I have to do with ants and the rock, or gold coins? I am in search of the highest knowledge. I am going to find a great Guru." His ego assured him that all was well.

They again set off. They came to a riverbank and waited for the boatman to ferry them across. As they waited, a huge fish rose out of the water, and the dervish said to Malik, "The fish is asking me to feed him some herbs, because a huge stone is stuck in his throat, and the herbs will help to remove it."

Malik said, "Oh dervish, what do we have to do with this fish and the stone in its throat? I'm in search of the highest knowledge. The boat has arrived. Let's get going." They sat in the boat and were ferried across the river. Night fell, and they lay down and went to sleep. Suddenly they heard a big commotion. They got up and saw the boatman coming, holding a beautiful ruby in his hand.

Malik asked the boatman, "Where did you find that?"

The boatman said, "Well, as I was coming to pick you up, I

heard the dervish say that the fish was suffering. So I fed it some herbs, and it coughed out this ruby."

This time Malik got angry. He looked at the dervish and said, "All this time you knew everything! You spoke to me in riddles, and I couldn't understand what you were saying. I missed everything."

The dervish patted him on the back. "Oh son," he said, "I told you, but you lacked the capacity to understand what I was telling you. You were seeking to learn from the greatest Guru, but you yourself have not taken the trouble to become even the smallest disciple! Until you become worthy through your own effort, you won't attain anything."

In that instant Malik realized his folly; he became a disciple and received initiation from the dervish.

So when you become more open and free from ego you are able to learn from everything and everyone.

Qualities to be Developed

A student of Yoga should endeavor to develop the following qualities in order to succeed in Yoga:

(a) Discrimination (*Viveka*): increasing understanding of what is true and what is essential.
(b) Dispassion (*Vairagya*): increasing strength of will to detach oneself from what is false and erroneous.
(c) Six-fold Virtues (*Shat-sampat*): 1. Serenity of mind, 2. Control of the senses, 3. Discarding of selfishness, 4. Endurance and patience, 5. Faith and firm conviction (*Shraddha*) 6. Balance of the mind.
(d) Intense aspiration for Self-realization (*Mumukshuttawa*).

He should not complain about his circumstances. He should rather convert everyday activities into a process of self-purification. There is no use in waiting for favorable circumstances. Can one wait for the waves to become still in order to enter the ocean? He has to make use of all circumstances as a means for spiritual practice.

The Disciple's Attitude toward his Guru

What should be the attitude of a disciple toward his Guru? Swami Vivekananda, the greatest among the disciples of Sri Ramakrishna, exhorts us on the basis of his own personal experience:

> "Find the teacher, serve him as a child, open your heart to his influence; see in him God manifested. Our attention should be fixed on the teacher as the highest manifestation of God, and as the power of attention concentrates there, the picture of the teacher as man will melt away, the frame will vanish, and the real God will be left there."

So we should approach the Guru with a spirit of veneration and devotion. We should not make the mistake of treating the Guru as equal. One of the prime attitudes that we should have towards our Master is humility. However, in America where there is the theoretical ideal that all are equal, Americans find it repugnant to place anyone above themselves. They may see a Guru as a mere (removed 1 word here) person. It is shocking to many of them to see disciples bowing their heads to the feet of a Guru. A true disciple sees the Guru as the true Spirit personified. He realizes that if he is to understand the teachings of the Guru, he must "stand under" him, and humble himself. He must not try to block the Guru's Spirit by his own coloration, nor by his own ego attachments. He should be a receptive vehicle for the Guru's teaching. He should realize that unless he empties himself, he cannot receive that teaching.

Lord Krishna describes the proper attitude of a disciple in the *Bhagavad Gita*:

> "Seek this (knowledge) by prostrating, questioning and serving. The wise Seers of the Truth will instruct you in that knowledge." (4. 34)

This verse points out the attitude of a student toward his Guru. *Firstly*, he should prostrate at the feet of the Guru. This is not to glorify him. It does not matter to him whether you fall at his feet or jump on his shoulders. By prostrating, the student annihilates his ego and manifests his humbleness. A good student goes to a Master to receive, to learn, not to exchange his ideas, nor to test the Master's knowledge. He should feel and tell the Guru, "I don't know anything. I have taken refuge in you. Please instruct me." It is only then that the real Guru-disciple relationship begins.

Secondly, the student should put his questions with sincerity to the Guru, in order to learn from him. "Ask, and it shall be given," says the *Bible*. Through questions only, can he get his doubts clarified. It is the duty of the teacher to convince the student.

Thirdly, he should serve the teacher. He should not get anything without giving everything he can. By serving the Guru he expresses his love and humbleness toward his Guru.

Here is a story

Once on a religious festival day, many devotees had gathered around Kabir (AD 1440-1518).[2] He just kept weaving and weaving. Finally one of the persons in the gathering stood up and asked, "Master, Who is a true disciple?" Kabir gave no answer. He kept on weaving. Again the person got up and asked, "Master, Who is a true disciple?" Kabir did not answer. He kept weaving. The person asked a third time.

At that point, Kabir called for his disciple, "Kamal come here." Kamal went to him and Kabir said, "Go and get some *prasad* (blessed food) for these people and when you bring it add some salt to it."

Kamal went and brought the sweet *Prasad* and added salt to it and distributed it to everyone. Everyone took the *prasad*, but it stayed on their palms!

After a while the same person stood up once again and asked, "Master, Who is a true disciple?" He really wanted to know.

Then Kabir said, "Kamal, come here." Kamal went to him. Kabir said, "I have lost my shuttle. Get a lantern and look for it." It was noon time, so there was good light. Kamal brought the lantern, found the shuttle and gave it to Kabir.

The same person got up and said, "Master, for the last time I am asking. Please tell me, 'Who is a true disciple?'"

Kabir said, "You still didn't get it? Kamal is a true disciple. He knows that salt is not added to sweet. Yet because I have asked him to do it, he did it. He also knows that he could just pick up the shuttle in the day light without a lantern. But he brought a lantern, because I told him to bring it. That is a true disciple: one who has such humility and obedience. In discipleship there is total annihilation of one's individuality.

Gratitude

The disciple should feel grateful to the Guru for the initiation and guidance that he receives from him. A great being called Sundardas described his experience of how he received initiation:

> "I went to my Master to receive initiation.
> What did he do? He gave me the fabulous Word.
> The moment he gave me the Word, it pierced my body.
> Just as the sun removes darkness,
> The Word removed my inner darkness, my delusion.
> I am so grateful for this.
> With my speech, with my mind,
> I offer my salutations to my Master."

When we receive something from others, gratitude should well up along with it. Often we lack gratitude in our lives. We receive so much from God, we receive so much from Nature, and we receive so much from other people, yet we are ungrateful. When we receive grace, we should be grateful for all that we receive. This gratitude may take many forms: love, reverence, compassion, a sense of indebtedness, and a way of life.

Devotion and Faith

Besides this, the student must have faith in the Guru, and have devotion to him. It is the faith and devotion that connect him with the Guru. The Guru will not force anything into the disciple. He will wait until the disciple asks for it, until he becomes ready. This is where devotion and faith come in. Unless the disciple has faith and devotion he cannot receive what the Guru has to give him. Devotion means that he puts his entire faith in the Guru. That faith becomes the connecting link between the Guru and the disciple. Once that link and communion is established, even if the Guru refuses to teach him, he will learn from the Guru, because of the power of faith. By his own faith the disciple will be able to understand what the Guru has in his mind; he need not even tell him.

But the devotion to the Guru is difficult. The student constantly has to prove himself to be a true disciple. Many incidents may occur and disturb his faith. However, with all that, if he feels, "He is my Master. I have full faith and implicit trust in him. I am not here to question or judge him; I am here to be trained by him," then that is deep faith.

If anyone has that faith, he is really fortunate. He is fit to get all because of his faith. He learns more from the Guru than even the Guru knows, because his faith itself acts as his Guru. Faith is God. That is why the *Bible* says that if one has faith the size of a mustard seed, one can move mountains. But if that is shaken, even if a disciple's Guru is God Himself, he will not receive anything, because there is no communion. Faith is the greatest virtue.

The real teaching or imparting of the true knowledge is not normally done with words. Words have their limitations, but by speaking through silence, in feeling, the disciple receives much more. The Guru knows what the disciple needs, and the disciple knows what the Guru wants to communicate – all without words.

The real learning, thus, comes more from feelings and observation than from words. How does a child understand the mother? She loves the child so much and the child reciprocates. Even if the mother is far away, she can easily feel when her baby is hungry. In the same way true devotion connects the student's heart with that of the Guru, and thus communion takes place between them even without words.

Guru Worship

In many cases the disciples worship the Guru as a deity. This practice is often misinterpreted as acts of investing the power of

God in a human being. Certainly there are some disciples who only worship the body, "the person" of the Guru; however, the true intent of (removed 1 word here) Guru worship is the disciple's devotion to the Guru's indwelling Spirit. It is that Spirit that the true disciple perceives in his worship of the Guru. In short, Guru worship is just a supplication to the Guru's spiritual essence.

Guru Service

Guru service is an important aspect of the disciple's practice. It is one part of Karma Yoga, selfless service. The disciple tries to become a perfect channel, an instrument for the Lord's work. Then he can feel the supreme power working through him.

Surrender to Guru

It is not enough for the disciple to simply present himself to the Guru; he must surrender to the Guru. If the Guru is to do any real work on him, he must give himself wholeheartedly. In India and other Eastern countries, it is not uncommon for the disciple upon meeting the teacher to kneel down and put his head at the Guru's feet. This symbolizes his humility. Another esoteric meaning of such worship is receiving the Guru's energy; for the head is thought of as the North Pole and the feet as the South Pole.

As a rule, humility and obedience are the hallmarks of a good disciple. When he surrenders, he is allowing himself to be worked on by the Guru or the Divine force around him. The disciple's self-surrender to the Guru and the Guru's grace are interrelated. The surrender draws down the Guru's grace and the Guru's grace makes the surrender complete. The Guru's grace works through the disciple's practice (*sadhana*). If the disciple sticks to the path tenaciously, it is the grace of the Guru. If he resists when temptations assail, it is the Guru's grace. If people receive him with love and reverence, it is the Guru's grace. If he gets all bodily wants, it is the Guru's grace. If he gets encouragement and strength when he is in despair and despondency, it is the Guru's grace. If he gets over body-consciousness and rests in his own Self and blissful nature, it is the grace of the Guru. Let the disciple feel the grace of the Guru at every step, and be sincere and truthful to him.

Introspection

An aspirant should cultivate introspection or self-analysis in preparation for spiritual training. He should analyze his motives and actions in order to learn about himself. Self-analysis helps him to identify his imperfections and to take appropriate steps to correct them. It helps him to develop an in-drawn attitude; it starts the process of detaching his mind from the objects of the world and placing it more securely on the Self. He should dedicate himself to the task. The first little taste of peace that he experiences becomes the impetus for further spiritual efforts. How have some achieved great things in science, exploration, arts, sports and other fields? It is all by determination and dedication. Therefore, self-surrender, introspection, sincere devotion and dedication are necessary for a disciple in preparing for receiving the Guru's grace and guidance.

REFERENCES

1. A poet-saint of Uttar Pradesh, Northern India. He composed bhajans and a number of poetical works on Vedanta.
2. A great poet-saint and mystic who lived in Benares, Northern India as a simple weaver. His influence was a powerful force in overcoming the fierce religious factionalism of the day. His poems describe the experience of the Self, the greatness of the Guru and the nature of true spirituality.

14

Mantra

Meaning

Mantra is a set of words or sounds such as *"Om Shanthi," "Hari Om," "Om Namasivaya"* or *"Jesus Amen,"* which is repeated with devotion and understanding for attaining spiritual illumination. The Sanskrit word *"mantra"* means "that which liberates the mind from bondage and suffering."

Mantra, when uttered with feeling, creates a specific psychic vibration. It is a key to unlock a potential treasure within. It awakens the inner latent powers of the soul, enabling a person to move towards God-realization.

Mantra is based on the understanding that sound is the medium of creation, the dynamic force of the Absolute. *Rig Veda* says, "In the beginning was Brahman, with whom was the word; and the word was truly the Supreme Brahman." Echoing this statement, almost exactly, the *Bible* says, "In the beginning was the Word and the Word was with God, and the Word was God." The "word" here means sound.

Om

Om is the mother of all *mantra*s. It is the source of all words and sounds. The sound *"Om"* or *"Aum"* is considered to be the vibration from which everything else is created. In the Jewish and Christian religions, it is *"Amen,"* and in the Islamic religion it is *"Ameen."*

The *Mandukya Upanishad,* which contains the essence of Vedanta, expounds the significance of *Om (Aum)*:

> "*Om,* which is the imperishable Brahman, is the universe. What was,
> What is and what shall be, and what is beyond is *Om*. (1)
> "All this is Brahman. *Om* is the symbol of Brahman. It stands for
> The manifested world and the unmanifested Absolute as well." (2)

This *Upanishad* further analyzes the word *Om (Aum)* and identifies each syllable with a particular stage of consciousness. Three syllables A, U and M respectively stand for waking, dream and deep sleep states, and the whole word *Om* represents the fourth state of *Turiya* (Super-consciousness).

Every language(do you mean "alphabet" here instead of language?) begins with the letter "A" or "Ah." It is pronounced by simply opening the mouth, without touching any part of the tongue or palate. "U" rolls from the very root up to the lips. "M" is produced by closing the lips. "*Om*" represents the entire process of sound. *Om* is the origin or seed from which all other sounds and words come. It is the most ancient word for God.

Om, the hum sound, is called *Pranava,* because it is connected with *prana,* the life force. *Prana* is the basic vibration that always exists whether it is manifesting or not. It is never-ending. Scientists find that when the manifested objects are reduced to their unmanifested state, they go back to the atomic vibration. Nobody can stop that vibration. Similarly, even without repeating the basic sound *Om,* it is always vibrating in you. It is the seed from which all other sounds manifest. The entire universe evolves from that and goes back into that again. So no other name can be more adequate to represent God. *Om* was not invented by anybody. God Himself manifested as *Om.*

The *Katha Upanishad* speaks of *Om* as the goal of spiritual endeavor and identifies it with the highest Reality:

> 'That goal which all the Vedas proclaim, which is implicit in all penance,
> Seeking which persons live as spiritual students, that I will tell you
> briefly:

It is *Om*. It is the supreme, immutable Brahman Itself.
One who knows It attains the highest fulfillment of life." (1. 2. 15,16)

The *Mundaka Upanishad* compares *Om* to a bow, and the purified and concentrated mind to an arrow, and exhorts us to shoot the arrow with intense alertness so that it may hit the target which is Brahman and becomes one with it. (2. 2. 4). *Om* stands for the revealed knowledge. It is the most sacred word and has been handed down from generation to generation from the most ancient times. It is used as a *mantra* in *Japa* and Meditation.

Patanjali, who has codified the system of Raja Yoga in his *Yoga Sutras,* refers to *Om a*s the symbol of *Isvara* or God. (1. 27). He further states that constant repetition of *Om* removes all obstacles to spiritual progress and leads to the awakening of the Self (1. 29).

What we call sound is only external physical vibrations. Subtler than audible sound are the electromagnetic waves, the vibrations of the ether. Subtler still are the thought waves. Thought itself is a manifestation of the eternal super-sensual Cosmic impulsiveness of the *Nada Brahman* (sound Brahman), the Cosmic Mind.

The *Nada Brahman* is not a mere theoretical concept. It can be experienced. It can be heard through the subtle mind. When the mind is purified and concentrated, one hears the eternal, subtle, cosmic vibrations as a continuous sound, the *anahata dhvani*. This sound can be heard only when the mind is calm and the spiritual current rises to a higher plane of consciousness.

The Origin of Mantras

Om represents the unmanifested Brahman, the Absolute Reality, and other *mantra*s (word symbols) represent the differentiated views of the same Being. The *mantra*s are not words or sounds formulated by scholars. They are sounds or vibrations experienced by the sages in the highest states of Illumination. That is, they are evolved out of the deepest spiritual perception of sages. When the sages transcend their own finite individuality and attain Cosmic Consciousness, they experience many things: bliss, Knowledge of Self, sounds and other things. The sounds or words so perceived by sages are called mystic sounds or *mantra*s. Some of the *mantra*s received today through initiation were experienced many thousand years ago and personally passed on through a chain of teachers and disciples.

A *mantra* is really a tiny seed planted within you. If it is used properly with the right attitude, the tiny seed grows into the whole experience and transforms your personality. Through proper repetition of the *mantra*, you can attain the highest Illumination.

A *mantra* that is read in a book or given to you by someone who is not a spiritual Guru will not have the effect of the same *mantra* received from a Guru. The Guru gives a *mantra* most suitable to your personality and outlook, like a doctor giving a suitable prescription. Receiving a *mantra* from a Guru is an initiation into that spiritual tradition from which the perception of the *mantra* arose. You are connected to that tradition through *mantra* initiation.

A *mantra* is not a secret sound. But as per the Tantric tradition, you may be asked to keep it a secret. This is done to avoid contamination by the doubts and criticisms of others, that would arise, if it were revealed to them. Such doubts dilute the effectiveness of the *mantra*.

Faith in Your Mantra

You should have absolute faith in your *mantra* and be absorbed in it. You should make the *mantra* completely "yourself" and "you" completely "it." When you and the *mantra* become one, you will also become one with the Personal God that the *mantra* represents. When you are ready, when the mind is properly cultivated and becomes absolutely pure, then even if you repeat it only once you will be transformed.

There was a disciple who had been studying under a sage for many years. One day while the Master was away, a man brought his little child who was dead. This gave him an opportunity to see the power of the *mantra*. He took a bowel of water, repeated the *mantra* three times, and sprinkled the water on the child; the child came back to life. The disciple was delighted; the father was overjoyed. He took his child home.

When the Master came back, the disciple told him the whole story. The Master got very angry. The disciple was really confused: it was such a great feat, yet the Master was upset.

He asked the Master, "Gurudev, please explain to me why you are so upset. Are you not happy that the child came back to life?"

The Master said, "That is not the point. I am sad that your faith in the *mantra* is so small. Why did you repeat it three times? Once, would have been enough!"

Such should be your faith in the *mantra*. The syllable that you repeat is charged with the awakened energy of the great Masters.

The Power of Mantra

The *mantra* or the Divine Name contains tremendous powers. Sri Chaitanya teaches us this truth: "Various are Thy names revealed by Thee into which Thou hast infused Thine own omnipotent powers, and no limitations of time for remembering those names are ordained by Thee." (*Siksastakam*, 2). The power of a *mantra* manifests itself only in qualified aspirants. Towards the end of his life, Sri Ramakrishna one day initiated his disciple Narendra with *"Rama"* mantra at the Cossipore garden-house. This produced a remarkable change in Narendra who was by nature endowed with great power of self-control. He entered into an ecstatic state, and went round and round the house repeating the name of the Lord *"Rama! Rama!"* in a high and excited voice. He had practically no outward consciousness and spent the whole night (removed 1 word) this way. When the Master was informed of this, he only said, "Let him be, he will come round in due course." Narendra came round after some hours. (Eastern and Western Disciples, *The Life of Swami Vivekananda,* Calcutta: Advaita Ashram, 1974, p.130).

The *mantra*s stand for the spiritual urges of human beings. Just as an ordinary word when heard or uttered can arouse in us a certain idea or desire, so also the *mantra*s can arouse in us our latent spiritual tendencies. They produce a powerful effect on our unconscious mind, and ultimately lead us to God-realization.

A beginner may not be able to understand the power of the *mantra*. But if he sincerely repeats it, he will gradually realize its power. He will be able to appreciate the power of the *mantra* through direct experience. To ordinary people a *Mantra* is a mere word or a formula. But to an advanced spiritual person it is a concentrated thought force of great power leading to profound spiritual experiences. The repetition of the *mantra* removes the impurities of the mind such as lust, anger and greed. Just as a cleaned mirror acquires the power of reflection, the purified mind acquires the capacity of reflecting the Divinity within. The recital of the *mantra* also destroys one's sins and brings everlasting peace, infinite bliss, prosperity and immortality. By the repetition of the *mantra*, all obstacles to God-vision disappear, and there dawns the Knowledge of God. A devotee, by constant repetition of his *mantra*, is tuned to the Cosmic Power. By that tuning he feels that force within him, imbibes Divine qualities, and attains the vision of his Personal Deity. Later on, transcending all sound vibrations, he reaches the supreme Spirit.

Practices

In *Japa* and Meditation, the repetition of a *mantra* is practiced for centering the mind. By the constant repetition of the *mantra* given by the Guru, you are merged in the sound vibration. Starting with mere repetition of the *mantra*, you are led spontaneously to Meditation upon its meaning and then to the realization of the truth it represents. The *mantra* also works on the level of your unconscious mind, controlling your moods and undesirable emotions like anger, greed, lust and so on. It purifies your mind, enabling you to attain a higher degree of Concentration.

The *mantra* is the most common object of Concentration in the Yoga of Meditation. The Yogi tries to focus the mind on the *mantra*. He repeats the *mantra* mentally and trains the mind to become one pointed. (For further details *see* chapter 24. Cultivation of Bhakti: Positive Practices – III.).

Types of Mantras

*Mantra*s are of various types. The most common types of *mantra*s in Hinduism are given below.

1. *Om Sri Ganeshaya Namah*

This is the *Mantra* of Lord Ganesha. It is repeated at the commencement of every undertaking or spiritual practice for the removal of obstacles, and for the attainment of success.

2. *Om Gurave Namah*

Mantra of spiritual preceptor or Guru, for invoking Divine Grace, spiritual knowledge and illumination.

3. *Om Sri Durgayai Namah*

Mantra of Goddess Durga, the destructive aspect of the Mother. It is recited for the removal of inimical forces and negative qualities, and all obstructions in both the material and spiritual fields.

4. *Om Srim Mahalakshmyai Namah*

Mantra of Goddess Lakshmi, Goddess of Wealth. It is effective for material and spiritual prosperity.

5. *Om Aim Saraswatyai Namah*

Mantra of Goddess Saraswati, Goddess of Learning. It is recited for the removal of inertia and other obstacles in the course of development of knowledge, and for the attainment of success.

6. *Om Namo Narayana*

Mantra of Lord Vishnu, recited for the attainment of material prosperity and liberation.

7. *Om Namo Bhgavate Vasudevaya*

Mantra of Lord Vishnu, but more specifically of His manifestation, Lord Krishna.

8. *Om Namahsivaya*
Mantra of Lord Shiva, for material and spiritual attainments.

9. *Om Sri Ramaya Namah*
Mantra of Lord Rama, highly effective for spiritual development, material success and liberation.

10. *Om Sri Rama, Jaya Rama, Jaya Jaya Rama*
Another *Mantra* of Lord Rama, for material and spiritual success.

11. *Sri Hanumate Namah*
Mantra of Lord Hanuman, the foremost devotee of Lord Rama, for physical and mental strength, the removal of obstacles, and for overcoming (used this word instead of encountering as it did not make sense in the context it was used) the more critical problems of life.

12. *Om Sri Krishnaya Namah*
Mantra of Lord Krishna, for material and spiritual attainments.

13. *Om Sri Saravanabhava*
Mantra of Lord Subramanya or Karthikeya, for material and spiritual attainments.

14. *Hare Rama Hare Rama*
Rama Rama Hare Hare
Hare Krishna Hare Krishna
Krishna Krishna Hare Hare
   Mahamantra, which combines the names of Rama and Krishna. It is effective for promoting a spiritual atmosphere within and without. It is suitable both for *Japa* as well as for singing in the form of Kirtana.

15. *Om Bhur Bhuvah swaha*
Tat Savitur Varenyam

Bhargo Devasya Dheemahi
Dhiyo yo nah Prachodayat

This is the *Gayatri Mantra*, for health, intelligence and illumination.
Its meaning:

Om, we meditate on that adorable effulgent God, who is the creator
of the physical, astral and causal planes, and who removes the
darkness of ignorance. May He enlighten our intellects.

16. *Om Tryambakam Yajamahe*
    Sugandhim Pushti Vardhanam
    Urvarukamiva Bandhanam
    Mrityor Muksheeya Ma Amritat

This is *Mahamrityunjaya Mantra*. It is repeated for long life,
removal of the malefic influences of the planets, and the attainment
of immortality. This *mantra* glorifies Lord Shiva. Its meaning is:

Om! We worship the All-seeing One,
The giver of fragrance and nourishment.
May He liberate us from the fear of death, and
Let us realize our Immortality.

17. *Om Soham*

Vedantic *Mantra*, meaning, "I am That." It is recited for enquiry,
contemplation and meditation leading to Self-realization.

18. *Om Aham Brahma Asmi*

Another Vedantic *Mantra*, meaning "I am Brahman," for enquiry,
contemplation and meditation leading to Self-realization.

19 . *Om Tat Twam Asmi*

Another Vedantic *Mantra*, meaning "Thou art That," for enquiry,
contemplation and meditation leading to Self-realization.

Christians recite such *mantra*s as *"Jesus Amen," "Hail
Mary," "Father in Heaven Amen"* and others in meditation and
prayers.

15

The Objects of Worship

Introduction

What are the objects that are worshipped? In all religions there are various grades of worship. The objects that are worshipped may be broadly classified into two categories: (1) God and (2) substitutes (*pratika*). God includes both unmanifested Supreme Consciousness or Impersonal God and manifested God or Personal God. The latter includes Incarnations, Chosen Ideals, words (Names), images and Gurus (Spiritual teachers).

The substitutes are of various forms: departed ancestors or departed friends or departed child, book worship, holy men, celestial beings (*deva*). Those who worship these substitutes are not worshipping God, but something, which is near. This worship cannot, says Swami Vivekananda, lead us on to salvation and freedom; it can only give us certain specific things for which we worship them. For example, one, who worships his departed ancestors or departed friends, may perhaps get certain powers or information from them; but liberation, the highest goal comes only by the worship of God Himself. The persons who do not understand anything higher may worship these substitutes, but after a long course of experience, when they will be ready to get freedom, they will, of their own accord, give up the substitutes.

Worshipping Departed Ancestors or Friends

Personal love for ancestors and friends is so strong that when they die, we cling on to their bodies, and we worship them. We must remember that the worship of these nearing-stages (*pratikas*) can never lead us to salvation. It is all right if they lead us on to a further stage, but the chances are mostly that we stick to these *pratikas* all our lives. We never come out of it or grow. That is the great danger of worshipping *pratikas*.

Book-worship

This is another form of strongest *pratika*. In every country the scripture comes to occupy the place of God. Even if the teachings of God as an Incarnate do not conform to the *Vedas* or *Bible*, people will not accept them. People do not want anything new, if it is not in the *Vedas* or the *Bible*. When you hear a new and striking thing, you are startled. The mind runs into ruts, and to take up a new idea is too much of a strain. Think of the mass of incongruities that the liberal preachers have poured into (removed 1 word) society. According to the Christian Scientists Jesus was a great Healer; according to the Spiritualists He was a great psychic; according to the Theosophists He was a Mahatma. All these have deduced from the same text.

Similarly, there is a text in the *Vedas* which says, "Existence (*Sat*) alone existed. O beloved, nothing else existed in the beginning." Many different meanings are given to the word *Sat* in this text. The Atomists say that the word means "atoms," and out of these atoms the world has been produced. The Naturalists say that it means "Nature." The Suniavadins say that it means "nothing." The Theists say that it means God; and the Advaitists say that it is the Absolute Existence. Yet all refer to the same scripture as their authority.

Some people cover the Holy Scriptures with gold and velvet cases, place them on the altar and worship them, but know nothing about what is inside.

These are the defects of book-worship. Though books have given strength to the religious faiths, they have produced more evil than good. They are responsible for many mischievous doctrines. All creeds come from books, but superficial knowledge and wrong understandings of books alone are responsible for persecution and fanaticism in the world.

Images

All over the world images are used in some form or other. Human beings, who want to worship an image naturally prefer to have it in the form of a human being rather than of an animal or any other form. God is omnipresent; He is manifesting Himself in every being., but for a human being God is only visible, recognizable in human form. When His light, His presence, His Spirit shines through the human form, then and then alone can people understand Him. So we have in the worship of God, forms and God-men. All religions have these in some form or other, bur still they want to fight with one another. As Swami Vivekananda says in his *Religion of Love*, 1970, one sect thinks a certain form is the right kind of image, and another thinks that is bad. The Christians think that when God came in the form of a dove it was all right, but if He comes in the form of a cow, as the Hindus think, it is wrong and superstitious. The Jews think that an image of a chest with two angels sitting on it and a book in it is all right, but that of a human being is awful. When the Muslims pray, they try to form a mental image of the temple with the Kaaba with the black stone in it and turn towards the West, but if you form an image in the form of a church, it is idolatry. This is the drawback of image-worship. (p. 66).

Yet images are necessary in the early stage of devotion. In trying to concentrate your mind, you require naturally to form images in your mind. You cannot help it. Except a human animal that never thinks of any religion, and the perfected sage who has passed through these stages, all others require some sort of ideal, outside or inside. We are prone to concretize. We are concretized or embodied souls and so we are bound to worship personal beings. It is easy to say, "Don't be personal; but the same person who says so is generally the most personal. His attachment for particular men and women is very strong; it does not leave him when they die, he wants to follow them beyond death. That is idolatry. Is it not better to have personal attachment to an image of Krishna or Buddha or Christ than to an ordinary man or woman?

However, the worship of images as images is not going to save you. You can worship anything, seeing God in it; forget the idol and see God there. This is beautifully portrayed by the ancient sage Thirumular in his Tamil scripture of *Thirumantiram* in the following verse:

> "*Marathai maraithathu mamatha yanai*
> *Marathin marainthathu mamatha yanai*
> *Parathai maraithathu parmutharputham*
> *Parathin marainthathu parmutharputhame.* (2290)
> Toy-elephant conceals wood.
> Recedes in wood toy-elephant.
> Five elements conceal Brahman (God).
> Recede in Brahman the elements."

A child looks at a toy elephant carved from wood. It sees only the elephant and it is not aware of the wood. A carpenter, on the other hand, looks at the quality of the wood and he is not bothered about the figure carved. Similarly we ignorant children are aware of the world constituted of the five elements and not of the Lord who has manifested as the world. A Jnani (wise), on the other hand, sees only God in everything. He is not aware of the material world.

So see God in everything that you may worship. God is in everything. You can worship a picture as God, but not God as the picture. In image worship you should superimpose on the image the God within you. Then you worship God in the image.

For an average person image-worship is indeed of immense help. But in the course of spiritual development, he has finally to outgrow all anthropomorphic images and comprehend the essence of God as indeterminable Being, ineffable Super-consciousness.

Divine Devotion frequently produces emotional attachment upon the chosen image. But that should not prompt the devotee to hate or downgrade other images. Such tendency is religious fanaticism. It hinders the unfolding of the cosmic vision of the Truth. All forms of worship are like different pathways to the same summit.

Rituals

Rituals are like ladders on which to climb up. If properly understood, they can help you to develop intense devotion to God. But they can also bind you, if you do them mechanically without knowing their meaning. That is why many people are discarding them. Many people are dissatisfied with their own religions because the disciplines, rituals and rules are presented without explanations. That will not work today. The present generation questions everything: "Why should we do it? What is the purpose behind it?" In fact, most of the rituals have lost their meaning because people do them mechanically.

Rituals are simply practical hints or external replicas of what should happen within you. If you take flowers to the temple or church, it is a way of expressing your love for God. If you bring fruits, you are bringing the fruits of your actions. In India devotees place two bananas before the altar. They signify that the results of both good actions and bad actions are offered to God. Some times there is a curtain or screen before the altar. When that is opened you are in the presence of God. This signifies the removal of the veil of your ego and your transcendence of it. The light that is waved before the altar signifies the removal of the darkness of your ignorance. These are the meanings of some of the rituals. (For further details *see* "Puja" in chapter 23. "Cultivation of Bhakti: Positive Practices II".)

The whole universe is a symbol, in and through which we are trying to grasp the things signified. This is what the lower human constitution needs, and we are bound to have it so. Rituals are like the mother's milk, which every baby takes at first. Just as the baby gives up mother's milk after a time, an aspirant must go to the things signified, to go beyond the material, and reach the spiritual. The spirit is the goal, and not the matter. Forms, images, candles, bells, rituals and all holy symbols are good and helpful to the growing plant of spirituality, thus far and no further. The symbols and rituals should not remain as permanent aids in your spiritual practice. You must rise above these and reach the higher stages of development, and see the Lord within you and in everything and every being in the universe. As Sri Krishna says in the *Bhagavad Gita,*

> "Whatever be the merits prescribed by the Vedas to accrue from
> the performance of sacrifices, austerities and offering of gifts,
> the Yogi goes beyond all these and attains to the Supreme State. (8. 28)
> "Only by single-minded devotion can I be known and seen
> in reality and also entered into, O Arjuna." (11. 54)

Word: The Power of the Divine Name

Many scriptures say that God created this universe out of "word" or "sound" (*Sabda-Brahma*). The whole universe is composed of names and forms. We believe that God is formless and nameless, but as we begin to think of Him, we give Him both name and form, as said by a Tamil saint: "We have given thousand holy names to One who has no form and no name." Our mind is like a calm lake and thoughts are like the waves upon it. Thoughts take names and forms. Thus the projection of anything in the mind cannot be without name and form. Every thought has a certain name and form. So the very fact of creation or (more appropriately) projection is eternally connected with name and form. This being so, it is natural to suppose that this universe is the outcome of mind, just as your body is the outcome of your idea. Your idea, as it were, made concrete and externalized. The universe is built on the same plan.

We know that body is a grosser and name is a finer state of the manifesting power called "thought." All these three—form, name and thought—are one. Wherever one is, the others are also there. Wherever name is, there is form and thought. As the universe is supposed to have built upon the same plan as our body, the universe also must have these three aspects of form, name and thought. The "thought" behind our body is called soul, and the "thought" behind the universe is called God. A particular person has a name. What is the name of that word—sound—that projected the universe? According to the ancient scriptures of India that word is *Om*. (For details *see* chapter 14. Mantra.) The ancient Egyptians also believed that *Om* stands for the name of the universe or God. It represents both of them.

God has so many names. All these names or words are sacred and have infinite power. Simply by repetition of these names with feeling we can attain a higher state of Bhakti and perfection. (Swami Vivekananda, *Religion of Love*, 1970, pp. 70-75)

The name of God is a verbal expression of the Supreme Being. Its glory cannot be established through reasoning or arguments. It can be experienced only through faith and belief. What a lot of joy its repetition brings! What a lot of power it infuses into the devotee! It destroys old sins, desires, whims, fancies, depressing moods, lust and other innate tendencies and raises him to the status of Divinity. It makes him realize his oneness with the Lord.

The prostitute Pingala was mysteriously transformed into a saintly lady by the power of repeating the name of Sri Rama. She obtained a lovely parrot as a present from a thief. The parrot was trained to utter the name "Sri Rama, Sri Rama." Pingala knew nothing of Sri Rama. She heard the sound Rama-Rama through the mouth of the parrot. It was melodious and charming. Pingala was very much attracted to it. She fixed her mind on the sound and mysteriously entered into *bhava Samadhi*. Such is the power of the name of the Lord.

Kabir, the poet-saint of Benares (Northern India) sent his son Kamal to the saint Tulsidas (sixteenth century). Tulsidas wrote the name of Sri Rama on a tulsi leaf and sprinkled the juice over many lepers. All were cured. Kamal was quite astonished. Then Kabir sent Kamal to the saint Surdas. Surdas asked Kamal to bring the corpse that was floating in the river. Surdas repeated Rama only once in one ear of the corpse and it was brought back to life. Kamal was filled with awe and wonder. Such is the power of God's name.

The continuous repetition of a name of God produces a deep-rooted impression in the mind of the repeater who ultimately attains

God-vision. There is an unfathomable depth, intense sweetness and charm in the Lord's name. It is all-bliss, and when it is chanted, the mind merges in its bliss, it loses its individual entity in the bliss, and it becomes one with the Bliss itself. God's name purifies the vicious lower self and elevates it to the sublimity of Cosmic Consciousness. "Repeat the name of the Lord with intense feeling and love from the very bottom of your hearts, with all your being, and realize the supreme bliss, peace and immortality," exhorts Sri Swami Sivananda, a great saint of this century.

The Chosen Ideal (Personal God)

Each one of us is born with a particular nature as a result of our past existence. Each one of us is an effect of our past. Each one's particular nature is called the "chosen way." This is the theory of *Ishtam* (The Chosen Ideal). This means each one of us sees God according to our own nature, and this vision is called our *Ishtam.* One person should not force another to worship what he worships. All attempts to make human beings to worship the same God have failed, because it is constitutionally impossible to do so. You cannot make a plant grow in a soil unsuited to it.

God has no form, no name. It is *Satchidananda*, Existence-Knowledge-Bliss. It is One without a second. It is all-pervading Consciousness. It is too much of an abstraction to be loved and worshipped. So devotion is directed towards the Absolute in Its personal aspect, the Personal God. The ideal of the Unmanifest is hard to attain for those who are identified with their bodies. So they choose the Personal God for devotion and worship.

The devotees who attain that state where there is neither knower nor knowable nor knowledge, where there is neither "I" nor "Thou," where there is neither subject nor object nor relation, become One with the Absolute. But for those who have not reached that state, there will remain the triune vision of the worshipper, the worshipped and the worship.

When the greatest devotee Prahlada forgot himself in meditation on the Lord, he found neither the universe nor its cause; all was to him one Infinite, undifferentiated by name and form. But as soon as he remembered that he was Prahlada, there was the universe before him, and with it the Lord of the universe. It was so with the blessed Gopis of Brindavan. So long as they had lost the sense of their own personal identity and individuality, they were all Krishnas, but when they began again to think of Him as the worshipped, then they were Gopis. (*Bhagavatam*).

A devotee can choose any name as his Chosen Ideal and worship through it, but he must take care not to hate or even criticize the Chosen Ideals of others, because all the names refer to the same God.

Bhakti Yoga, therefore, lays on the imperative command not to hate or deny any one of the various Ideals that lead to salvation. Yet the growing young plant must be hedged round to protect it until it has grown into a tree. The tender plant of spirituality will die if exposed too early to the constant change of ideas and ideals. Many people, in the name of religious liberalism may be jumping from one Ideal to another. In the words of Swami Vivekananda "Religion is with those people a sort of intellectual opium-eating, and there it ends."

"There is another sort of man," says Bhagavan Ramakrishna, "who is like the pearl-oyster of the story. The pearl oyster leaves its bed at the bottom of the sea and comes up to the surface to catch the rainwater when the star Svati is in the ascendant. It floats above on the surface of the sea with its shell wide open until it has succeeded in catching a drop of the rain-water, and then it dives deep down to its sea-bed and there rests until it has succeeded in fashioning a beautiful pearl out of that raindrop." This is indeed the most poetical and forcible way in which the theory of devotion to the Chosen Ideal has ever been put.

Devotion to one Ideal is absolutely necessary for the beginner. He must say with Hanuman in the *Ramayana*, one of the great Epics of India: "Though I know that the Lord of Rama and the Lord of Janaki are both manifestations of the same Supreme Being, yet my All in all is the lotus-eyed Rama." "Then if the devotional aspirant is sincere, out of this little seed will grow a gigantic tree, like the Indian banyan, sending out branch after branch and root after root to all sides, till it covers the entire field of religion. Thus will the true devotee realize that He who was his own ideal in life is worshipped in all ideals, by all sects, under all names and through all forms." (Swami Vivekananda, *Karma Yoga and Bhakti Yoga*, 1982, p. 161)

16

Attitudes (Bhavas) of Devotees

Introduction

It is impossible to express the nature of Divine Love in human languages. Even the highest imagination is incapable of comprehending its infinite perfection and beauty. Nevertheless, the followers of the Path of Devotion to God have all along had to use our limited human language to comprehend and to define their own ideal of love. Human love, in all its varied forms, has been used to typify this inexpressible Divine Love. We can think of God only in our own human way; we can express the Absolute only in our relative language. Therefore, devotees make use of all the common terms associated with the human love, in relation to God.

An important factor in Bhakti Yoga is the inner attitude (*Bhava*) that expresses a particular relationship with God. It is purely a mental state. There are six kinds of attitudes in the practice of Bhakti Yoga. They are calmness (*santa*), servanthood (*dasya*), friendship (*sakhya*), childhood (*vatsalya*), parenthood (*tata*) and sweetheart relationship (*madhurya*). A verse in the *Prapanna Gita* beautifully portrays a true devotee's attitude towards the Lord:

"O Supreme Lord, Thou art my father and mother
Thou art my relatives and friend
Thou art my knowledge and wealth
Thou art my all in all." (28)

Calmness (Santa)

One form of attitude towards God is calmness of mind. In this attitude, the mind of the devotee is filled with divine knowledge and it is emotionless; it remains peaceful without the fire of love or madness of intensely active love. It can be practiced only by Yogis and rishis who are highly developed and whose emotions have been stilled. Bhisma, the grandfather of the Pandavas, had all his emotions under control. He was a *santa* devotee. The *santa* devotee is always undisturbed, peaceful and tranquil. He rests peacefully in God and appears like a lake without waves.

The *santa* devotee is indifferent to worldly things and is entirely dispassionate. The *santa bhava* is developed through the study of *Upanishad*s and such other scriptures, through living the life of seclusion, and association with persons of similar nature. This bhava is also aroused by intuitive flashes of Divinity.

Servanthood (Dasya)

In this form of devotion, the devotee considers himself as a servant of the Lord. He takes God to be his Master. He craves for God's protection. He feels that he has no power. As he is a servant, the master would take care of him. He prays to God to save him. In this way he is engrossed in thoughts of God alone, until finally he loses himself in God. In his prayer Sri Chaitanya sings:

> "A drowning man in this world's fearful ocean
> Is Thy servant, O Sweet One, in Thy mercy
> Consider him as dust beneath Thy feet."

Serving and worshipping the Deities in temples, sweeping the temples, meditating on God and mentally serving Him, serving the saints and sages, serving the devotees of God, serving poor and sick people who are forms of God is included in *Dasya* Bhakti.

The purpose behind *Dasya* Bhakti is to be ever with God in order to offer service to Him and win His Divine Grace and attain thereby immortality. It means total dedication of the self to God. Lakshmana, Hanuman, Angada and others of the Epic *Ramayana* were great *Dasya* devotees of Lord Rama. Lakshmana could not even speak to Rama at times in his extreme devotion to Rama. Hanuman was a towering example of Divine service to the Lord. He spent his whole life in serving Lord Rama. Angada did not want to leave Rama even at the latter's request. These devotees were of an exceptional type and it is very difficult to develop such devotion. Total surrender is the ideal of *Dasya* Bhakti. The *Dasya* devotee loses nothing. He gains everything through the service of God. Several *Nayanmars* of Southern India, realized devotees of Lord Shiva, were *Dasya* devotees.

Friendship (Sakhya)

In *Sakhya Bhava*, God is taken to be a friend. Just as there is equality between friends, so equal love flows between the devotee and the Lord. God becomes our friend to whom we may reveal the inner most secrets of our hearts. God is viewed here as our playmate. Just as children play their games, so is the beloved Lord Himself playing in this universe. He is perfect. He does not want anything. Why should He create this universe? It is all His play. God is playing with us and we are playing with Him.

It is only when we forget that it is all play, it is only then that misery and sorrows come. But as soon as we realize that the changing world is a mere stage on which we are playing, at once misery ceases for us.

The cowboys of Brindavan are examples of *Sakhya* devotees. Krishna was their beloved friend. He played with them. Arjuna was also a friend of Krishna.

There is great love between true friends. Such a love is turned into spiritual love. The devotee considers all as God. He treats every being as his own friend. There is no selfishness, no hatred, and no separateness in him. He becomes one with all in feeling. He loses all, for all are his friends. All is God. And God is his supreme friend. He is always satisfied with what is ordained by God. God is the inner most and dearest of friends. All friends may desert a person, but God will never desert His devotees.

Childhood (Vatsalya)

In this type of relationship the devotee loves God as his child. This attitude enables the devotee to detach all ideas of power and awe from the concept of God. There should be no awe in love. The father and mother cannot think of asking any favor from the child. The child is always a receiver, and out of love for it, the parents will sacrifice everything.

This idea of loving God as a child comes into existence naturally among those sects, which believe in the incarnation of God. For the Muslims it is impossible to have this idea of God as a child. But the Hindus and Christians can realize it easily; they have the Baby Krishna and the Baby Jesus respectively.

Many women in India throughout the ages have looked upon themselves as Krishna's mother. The outstanding example in the present age was Aghormani Devi, a woman disciple of Sri Ramakrishna. She became well known as "Gopal's Mother." Widowed at an early age, she had dedicated herself completely to spiritual pursuits. She worshipped Gopala, the Baby Krishna, regarding Him as her own child. She worshipped Gopala, cooking for Him, feeding Him, bathing Him, and putting Him to bed. For forty years she had lived on the bank of the Ganges in a small hut. At the age of sixty, in 1884, she visited Sri Ramakrishna at Dakshineswar. During the second visit, as soon as the Master saw her, he said, "Oh, you have come! Give me something to eat." With great hesitation she gave him some ordinary sweets that she had purchased for him on the way. The Master ate them with relish and asked her to bring him simple curries or sweets prepared by her own hands. Gopal Ma thought him a queer kind of monk, for, instead of talking of God, he always asked for food. She did not want to visit him again, but an irresistible attraction brought her back to the temple garden. She carried with her some simple curries that she had cooked herself.

One early morning at three o'clock, about a year later, Gopal Ma was about to finish her daily devotion, when she was startled to find Sri Ramakrishna sitting on her left, with his right hand clenched like the hand of the image of Gopala. She was amazed and caught hold of the hand, whereupon the figure vanished and in its place appeared the real Gopala, her Ideal Deity. She cried aloud with joy. Gopala begged her for butter. She pleaded her poverty and gave Him some dry coconut candies. Gopala sat on her lap, snatched her rosary, jumped on her shoulders, and moved all about the room. As soon as the day broke she hastened to Dakshineswar like an insane woman. Of course Gopala accompanied her, resting His head on her shoulder. She clearly saw His tiny ruddy feet hanging over her breast. She entered Sri Ramakrishna's room. The Master had fallen into *Samadhi*. Like a child, he sat on her lap, and she began to feed him with delicacies. After some time he regained consciousness and returned to his bed. But the mind of Gopala's Mother was still roaming in another plane. She was steeped in bliss. She saw Gopala frequently entering into the Master's body and again coming out of it. When she returned to her hut, she was still in a dazed condition. Gopala accompanied her.

She spent about two months in uninterrupted communion with God, the Baby Gopala never leaving her for a moment. Then the intensity of her vision was lessened; had it not been, her body would have perished. The Master spoke highly of her exalted spiritual condition and said that such vision of God was a rare thing for ordinary mortals.

Sweetheart Relationship (Madhurya)

The highest expression of Divine Love is that felt in this relationship. This is the attitude of the lover and beloved. Here the devotee takes the Lord to be his Beloved. In this sweet representation of Divine Love, God is our husband. Mirabai loved Krishna as her husband. She was a queen, married to a Rajput king, but she renounced her husband and kingdom and went to Brindavan. At that time another saint, a disciple of Sri Chaitanya, lived in Brindavan. Mirabai wished to visit the holy saint. But the saint first refused saying that he did not want to see any woman. To that reply, Mirabai retorted saying that she did not know that any man lived in Brindavan except Sri Krishna, her Beloved. Hearing this, the saint came running to meet this great saint Mirabai. The love of the Gopis, the milkmaids of Brindavan for Krishna is a well-known example of this sweet relationship. Speaking of them, Swami Vivekananda says:

> "When the madness of love comes in your brain, when you understand the blessed Gopis, then you will understand what love is. When the whole world will vanish, when all other considerations will have died out, when you will become pure-hearted with no other aim, nor even the search for Truth, then and then alone will come to you the madness of that love, the strength and the power of that infinite love which the Gopis had, that Love for love's sake. That is the goal. When you have got that you have got everything."

Sri Ramakrishna tells us that the bliss derived from loving God as the beloved is eternal and infinite. The zenith of love is reached between the lover and the beloved. The highest intensity of bliss is attained in *madurya bhava*. It is not an erotic craving. It is pure love that cannot be understood by lustful persons.

In the "*Song of Solomon*," we find Divine Love described as the "Sweet" relationship:

> "Let Him kiss me with the kisses of his mouth;
> For Thy love is better than wine.
> Because of the savor of Thy good ointment,
> Thy name is as ointment poured forth,
> Therefore do the virgins love Thee."

In "*The Dark Night*," a poem by St. John of the Cross, we read how the lover is brought to the Beloved, and how a mystic marriage takes place:

> Upon my flowery breast wholly for Him
> And save Himself for none, there did I give
> Sweet rest to my Beloved one.

The bliss of *madhurya bhava* is absolutely different from conjugality of earthly experience. Earthly conjugality is purely selfish and is undertaken only because it gives pleasure to one's own self. But the love for God is not for the sake of the devotee. It gives pleasure to God. Strong selfishness is the root of worldly passion; but Divine Love is the outcome of loss of egoism. No earthly affection, however perfect it may be, can lead one to the supreme joy of Divine communion. The nature of Divine Love cannot be understood, so long as man is only a man and woman is only a woman. The austere transformation of the human into the Divine is the beginning of true love for God. In this love the individual consciousness is lost in God-consciousness.

These are the different aspects of Divine Love represented in human terms and in human language. But it does not mean that there are only these six aspects of Divine Love. There are possibilities of as many aspects as there are human relationships. Narada points out that even if externally they appear as different, they are all manifestation of the One Divine Love. The difference in attitudes can only be attributed to the differences in tastes, preferences, and predilections arising out of the past impressions of the individuals, or to some inscrutable Divine purpose to be worked out only in particular ways.

The Gopis of Brindavan were attracted by Krishna's beauty; Uddhava, Sudama, Arjuna and Kuchela had the attitude of friendship; Rukmini and Sathyabhama loved Him as a husband; and Kausalya, Devaki and Yasoda loved Krishna as their son. Ambarisa spent his whole life in worship, Prahlada in remembrance, and Hanuman in service. By this specification it does not mean that other attitudes were not found in their lives; it means only that each one is characterized by a predominant attitude. The different attitudes may be found in the same person at different times as in Sri Ramakrishna. In reality, when love for God arises in the heart, in any aspect, that love is overwhelming and, so intense, that the devotee forgets this world and all earthly ties.

All the different kinds of love, which we see in the world, have God as the one goal. But unfortunately a human being does not know the infinite Ocean into which this mighty river of love is constantly flowing, and so, foolishly, he often directs it to human beings. If the love in you is bestowed blindly on any human being, it will bring, sooner or later, pain and sorrow as the result. Our love must, therefore, be bestowed on the Highest One, the eternal Being. Love must reach its destination, the infinite Ocean of Love, just as all rivers flow into the ocean. God is the goal of all our passions and emotions. They are meant for Him. God is the beloved. Whom else can this heart love?

When God, the highest ideal of love is reached, philosophy, freedom, and salvation – all are thrown away.

When a devoted devotee realizes God he is liberated and attains perfect freedom. Yet though he knows that he is now One with God, he longs to continue to taste the Divine Love. "I want to taste sugar. I may know that I am He yet I will take myself away from Him and become separate, so that I may enjoy the Beloved." Love for love's sake is the devotee's highest enjoyment.

In the words of Swami Vivekananda,

"We . . . . begin as dualists in the religion of love. God is to us a Separate Being, and we feel ourselves to be separate beings also. Love then comes between, and man begins to approach God; and God also comes nearer and nearer to man. Man takes up all the various relationships of life – such as father, mother, son, friend, master, and lover – and projects them on his ideal of love in his God. To him God exists as all these. And the last point of his progress is reached when he feels that he has become absolutely merged in the object of his worship. . .He realizes at last the beautiful and inspiring truth that love, the lover and the Beloved are one."

(*Karma Yoga and Bhakti Yoga*, 1982, p. 208)

17

Nine Modes of Bhakti

Devotion to God is developed in different ways. It is the supreme attachment to God with intense inner feeling. It is an exclusive love for God. The devotee's external qualities do not matter at all. He may be illiterate, ugly or poor. That is immaterial. What is important is his intense devotion. God is satisfied only with intense devotion, not with other things. How is devotion to God expressed? In the *Bhagavatam* and the *Vishnu Purana* it is stated that there are nine modes or forms of Bhakti for expressing our devotion to God. They are:

1. *Sravana*: listening to the Lord's glories
2. *Kirtan*: singing of His glories
3. *Smarana*: remembrance of His Name and Presence
4. *Padasevana*: service of His feet
5. *Archana*: offering of flowers in worship
6. *Vandana*: prostration to the Lord
7. *Dasya*: cultivating the inner attitude of servant to God.
8. *Sakhya*: cultivating the attitude of a friend
9. *Atma Nivedana*: complete self-surrender to the Lord

We can practice one or any combination of the above modes of Bhakti according to our taste and temperament. Through that practice, we will attain Divine illumination.

Sravana

Sravana is listening to God's plays (*lilas*), His virtues, glories, and stories connected with His name and form. The devotee gets absorbed in the Divine stories, and his mind merges in the thought of the Divine. It cannot think of anything else. The devotee's mind is no longer attracted by the world. He remembers God only even in his dreams.

The devotee should sit before a learned teacher or saint and listen to the Divine stories with a sincere heart, without the sense of criticism or fault finding. He should try his best to live up to the ideals preached in the scriptures. Lord Krishna says in the *Bhagavad Gita*,

> "Know that by long prostration, by service and enquiry,
> The wise who knows the Truth shall instruct thee in that wisdom." (4. 34)

The *Bhagavatam* describes the glorious pastimes and sport of Lord Krishna and it is a favorite scripture of all devotees. In it is told that the best virtue in this world is to hear the Lord's glories, for by hearing them one attains to the Divine abode.

In *Sravana Bhakti*, the company of the wise plays a very important role. Mere reading by oneself is not of much use. Doubts will arise and only the wise can help you to clear them by giving correct explanations. Sri Sankaracharya says: "The company of the wise even for a moment becomes a boat to cross the ocean of worldliness."

King Parikshit attained liberation merely by listening to the glorious pastimes of Lord Krishna from the lips of the great saint Sukadev. His heart became purified within seven days.

Kirtan

Kirtan is singing the Lord's glories. It gives joy to the mind and at the same time it purifies the heart. It is suited to all without any distinction. It is the easiest of all modes of approach to God. It is the prescribed method for this age.

The devotee is thrilled with pure Divine emotion. He loses himself in the love of God. He weeps when thinking of the glory of God. His voice gets choked. He goes into a state of Divine ecstasy. He constantly chants the Name of the Lord. Wherever he goes, he sings and glorifies the Lord. He asks others to join him in singing. He sings and dances in ecstasy. Music melts the heart of even the stonehearted persons. Its use in devotional *Kirtan* gives blissful ecstasy to both the singer and the listeners. It sublimates their emotions.

Divine beings like Narada, Valmiki, Sukadev, Chaitanya, Nanak, Tulsidas and Surdas attained God by singing *Kirtan* alone.

It is written in the *Bhagavata Mahatmya* that when Sukadev was singing *Kirtans,* the gods themselves came down from heaven and took part with their musical instruments. Narada played his Veena, Indra played the mritanga (drum), Prahlada played his cymbals and danced, Lord Shiva did the same, and Lord Vishnu was also present. They were thrilled by the occasional dance of Sukadev himself.

During the last century there lived in Southern India a great singer-saint called Tyagaraja. He was a great devotee of the Lord Rama. He was considered the most dedicated person because he lived in God always. He saw God, he spoke to Him, and he was so close to the Lord that he even teased Him.

His songs are very beautiful. Such songs can not be composed just from the skill. They flowed from his intense devotion. In one of his songs, he says, "Hey, Rama! How tasty and sweet is Your name, Your name is sweeter than You." Yes, the names of God are sweeter than God Himself. You can not taste God, but you can taste His name. If you just repeat the name, within a few seconds, you begin to experience the indescribable bliss, which melts you.

The Lord cannot give Himself directly. It is like millions of kilowatts of electric power that cannot come to your home directly. If by any chance it did, you and your home would be destroyed. So out of His greatness and His love to serve us, He limits Himself and comes down through the word, the holy name, which we can repeat and taste the bliss. The more we chant or repeat it, and the more we praise His name, the more we go into ecstasy. That is why Tyagaraja says, "Rama, Your name is so sweet."

By singing His glory, by constantly repeating His name, we develop devotion expressing it more. We live in God and God lives in us. The vibrations of the chants and songs elevate us to another plane. Every cell of our bodies vibrates on that Divine level. We forget our bodies. We forget ourselves.

Chanting or repeating God's name transforms our emotion into Devotion. Through Divine Love we become more loveable. Singing the glory of the Lord or chanting does not require any rigid discipline. Through this practice we will get self-discipline, and in due course we will attain God-realization.

Smarana

Smarana is remembrance of the Lord at all times. It is continuous memory of the Name and form of the Lord. The mind is not allowed to think of any object of the world, but is kept engrossed in the thought of the glorious Lord. The unbroken remembrance of the Lord is the outcome of intense devotion. The silent repetition of the Name of the Lord, listening to stories of God, and Kirtan are the modes through which this is achieved. The company of holy people is a great help in keeping engrossed in the thought of God. Constant remembrance of God is the fruit of almost all other methods of spiritual practice.

God is to be remembered at all times. From the moment of waking up until we are overpowered by sleep at night, we should try to remember God, as this alone destroys all worldly habits and turns the mind away from sense-objects.

The miser does not forget his wealth even when he is busy with the duties. The cow does not forget its calf even while grazing. Similarly we should practice constant remembrance of God even when we are engaged in our respective duties. While the hands are on the work, let the mind rest on God.

Smarana serves as an effective whip to discipline the extrovert mind. The mind usually turns outward. It seeks pleasures outside in sense-objects. It wants excitement. *Smarana* puts a stop to this drama and disciplines the mind. Intense remembrance of God brings the mind under control. It no longer desires to run outward. It becomes meditative and can easily be concentrated.

Remembrance of God is not an easy method of practice. It is swimming against the current of the natural tendency of the mind. Conscious and continuous effort is required until remembrance becomes a habit. It is a rewarding effort. When we attain success in this effort we will attain God. Lord Krishna says in *Bhagavad Gita*,

"He who, fixing the mind on Me (Lord), constantly remembers Me,
Easily attains Me and is ever united with Me." (8. 14)

Remembrance of the Lord has given liberation even to those who remembered Him through hatred. Haters of the Lord like Kamsa and Sisupala had attained the state of liberation through *vaira* bhakti (devotion of enmity).

Prahlada practiced the remembrance of God even under trying conditions. His cruel father punished him in all possible ways, because he did not like Prahlada's devotion to the Lord. But the devoted Prahlada crossed over all the troubles and obtained the supreme Grace of the Lord. He was lost in the consciousness of God. *Bhagavatam* gives a beautiful description of Prahlada's intense devotion.

Padasevana

Padasevana means serving the feet of the Lord. It is not actually possible to do so, for the Lord is not visible to the physical eyes. However we can serve the image of God in idols or, better still, serve the whole humanity as God's manifestation. The great saint Thirumular of Southern India beautifully describes how service to humanity is service to the Lord, in his sacred text of *Tirumantiram*:

Padamada koyil Bhagavarkon treyil
Nadamada koyil nambark ankaka
Nadamada koyil nambarkon treyil
Padamada koyil Bhagavarka thame. (1857)
Offering given to the towered temple
Reaches not the noble walking temples.
Offering given to the noble walking temples
Reaches surely the Lord in the towered temple.

The offerings of such things as fruits, milk, honey, cooked food, etc., given to the Lord in the high-towered temple do not reach the Lord's noble devotees, who are walking temples. But the things offered to the devotees, who are the walking temples, surely reaches the Lord in the high-towered temple. Thus service to humanity is serving the feet of the Lord. Serving the sick and the poor, and service to other living creatures are also service of God.

Padasevana can also be performed as a ritual by doing formal worship of idols in temples, or to a mental image of God. The devotee observes the sacred feet of the Lord with deep feeling and humility, worshipping and serving them, bathing them and sipping the sacred water used to bath them. The holy sandals of the Lord are worshipped. The dust of the Lord's feet is applied on the forehead.

Holy shrines are respected. Centers of pilgrimages are considered sacred. The holy rivers are worshipped, regarding them as flowing from the feet of the Lord. Bathing in the holy waters and drinking the holy water are also forms of *padasevana*. Such devotion destroys worldly attachments and allows the mind to think exclusively of God.

The story of Bharata, Sri Rama's brother, in the *Ramayana* is a classical example for *Padasevana*. He asked for Sri Rama's sandals as an object of worship, and placed them on the throne to remind him that it was really Sri Rama who governed the kingdom, and that he was acting only as His instrument. He was indeed a great devotee of Lord Rama.

Bhagavatam says,

> "He who has taken shelter in the feet of the Lord, which is the refuge of saints and full of blessedness, for him the ocean of worldly existence becomes as shallow as the hoof-mark of a calf. The Supreme State is already attained by him. Worldly miseries and difficulties do not appear before him."

Archana

Archana means offering flowers in the worship of the Lord. An image of the Lord or a picture or a mental form is worshipped. God is omnipresent as the Cosmic Consciousness, but due to our limitations we try to perceive and worship the Cosmic Presence through an object. We should, however, always remember that God is not really limited to an object or image. It is important to have full concentration on the form of God during the worship. We should not think of worldly matters during the worship. The object of the worship is to destroy worldly attachments and develop pure love for God.

The *prasad* offered to God must be eaten with reverence. There are cases where devotees with burning love for God offered food to God and He actually appeared in supreme form and partook of the food offered to Him. These are rare cases, as true devotees are rare indeed.

Offerings to God need not be expensive. Gajendra offered Him only a flower from a tank, Draupadi offered a vegetable leaf, and Sabari some wild plums. Yet the Lord was pleased. It is the feeling of love that really matters, not the material value of what is offered. God is pleased even with leaves and water..

Advanced devotees may worship in any way they choose. They may have a mental picture of Him and offer imaginary flowers and garlands. The purpose of such mental worship is to please the Lord, to purify the heart through giving up the ego, and to cultivate intense love for God.

Serving the poor and worshipping saints is worship of the cosmic form of the Lord. God is everything. The universe and everything in it are God's manifestation. With this awareness if the devotee serves others, it is the highest form of worship. (*See* also *23. Cultivation of Bhakti: Positive Practices II*, Below)

Vandana

Vandana is prayer and prostration. It is humble prostration before the Lord with faith and reverence. Prostration to all beings, regarding them as the forms of the Lord, is supreme devotion. Lord Krishna gave this advice to Uddhava:

> "Giving no attention to those who laugh in ridicule, forgetting the body and being

insensible to shame, one should prostrate and bow down to all beings, even
  to a dog, an ass and a cow. All is Myself, nothing but Myself."
  (*Mahabharatam*)

When Arjuna beheld the Cosmic Vision of Lord Krishna he offered prostration and salutation to Him in a wonderful manner:

"Salutation to You from the front, salutation to You from behind,
  Salutation to You from every side! O All! Immeasurable in strength,
  You pervade All. You are all!" (*Bhagavad Gita*, 11. 40)

The purpose of *Vandana Bhakti* is to destroy the ego-sense, and to become humble. In the highest form of *Vandana*, there is perfect humility. In this state the devotee feels oneness with all creation. He regards even a worm with great reverence. He no longer has any feeling of superiority or inferiority. Divine Grace descends on such egoless devotees. The Divine Master, Swami Sivananda, used to prostrate to all without any reserve. He used to prostrate mentally to dogs, stones and bricks.

Dasya Bhakti

Dasya Bhakti is the Love of God through servant-sentiment. It means serving God and His creation, merging one's will with God's Will, with an attitude of servant-hood.

As mentioned in the previous chapter under the subtitle "Servant-hood," worshipping the idols in temples, sweeping the floor of temples, meditating on God and mentally serving Him, serving saints, serving the devotees of God, serving the poor and sick who are forms of God–all these are forms of *Dasya Bhakti*.

The purpose of *Dasya Bhakti* is to become in tune with God's Will. By being ready to serve Him through the needy people, the devotee destroys his self-will. He thus wins God's Grace and attains immortality.

Mother Teresa (1910-1997) of the Twentieth Century was an illustrious example for serving the Lord in the poorest of poor. She was born as Agnes Gonxha Bojaxhiu in Skopje, Macedonia. At the age of eighteen, she left home for Ireland to become a nun. When she was leaving home, her mother told her: "You go, put your hand in Jesus' hand, and walk alone with Him."

She became a nun in the Sisters of Loreto Teaching Order and took the name Sister Teresa.

She arrived in India in 1929 to begin her novitiate at the Loreto Convent in Darjeeling, in the foothills of the Himalaya Mountains. She took her final vow in 1937. At that time she was teaching at the Order's St. Mary's High School in Calcutta. She was named principal of the school, but she apparently could not shake from her mind the scenes of poverty and misery visible just outside the Convent wall. She would go out into the slums. In 1946, she fell ill with suspected tuberculosis and was sent to Darjeeling to recuperate.

It was in the train she heard the call to give up all and follow the Lord to the slums to serve Him among the poorest of the poor. Two years later, Pope Pius XII permitted her to leave the Order. She became an Indian citizen and received training in medicine to prepare for her new mission. She became well-known by teaching Calcutta slum children where families could not afford to send them to school. The children called her Mother. One day in 1948, she found a woman 'half-eaten up by maggots and rats' lying in the street in front of a Calcutta hospital, and sat with her until she died.

So Mother Teresa appealed to authorities for a building where the poor could die in dignity, and they gave her a hostel used by pilgrims' next to the temple of Kali. She started there the Nirmal Hriday (Pure Heart) Clinic. She and a small group of nuns roamed Calcutta's slums helping destitute people lying in the gutters. The Clinic remained the center of Mother Teresa's growing charity and the place she called Home.

Her work outside the convent was so successful that former students of hers soon joined her new Order as Missionaries of the Charity Home. Rome formerly approved it in 1950. Mother Teresa later described the purpose of the work: "We are missionaries to bring God's love and compassion into the world of today. That very fact, that we bring love to the hungry and the naked and the homeless and the dying, that we take care of them, is because Jesus has said to do so."

Her life changed when the British Broadcasting Corporation made a documentary on her in 1969. She became known as "the Saint of the Gutters." Volunteers and donations began pouring in.

Mother Teresa was awarded the Nobel Peace Prize in 1979. The Chairman of the Nobel Peace Prize Committee said, "Mother Teresa stands out in a very positive way, as an example of true self sacrifice in humanitarian work." She became a symbol to the world.

"Eileen Egan, her biographer, once said to her, she views herself as a vessel of good and of God, she sees Jesus in every person, even in the person who presents the most distressing disguise of Jesus."

Mother Teresa taught the world compassion through her deeds and demonstrated faith through her works. The Order she founded now includes more than 4,500 nuns, 500 brothers, and a handful of priests, as well as tens of thousands of lay volunteers. Her faith kept her working tirelessly until she had built, at the time of her death, a world wide network of orphanages, homes for the poor, AIDS hospices and soup kitchens. In every step she took, Mother Teresa saw a spiritual plan unfolding.

Mother Teresa became an icon of austere self-denial. She was an embodiment of love, compassion, devotion, dedicated selfless service, an exemplary Karma Yogi, God in human form. She saw God in the poor, in the destitute, in the sick and in the homeless, and served God through them.

Sakhya Bhakti

Sakhya Bhakti is the cultivation of friendship sentiment with God. The *Mahabharatam* says, "Oh, how wonderful is the fortune of the people of Vraja, of cowherd Nanda, whose dear friend is the perfect eternal Brahman of absolute Bliss!"

The *Sakhya Devotee* is always in the company of the Lord, treating Him as his close friend, and loves Him as his own Self. With this attitude, the devotee joyfully takes up any work of the Lord, and is concerned solely with the interests of his friend, the Lord. He considers his very life and all that he possesses fruitful only if they are useful in the service of God, for he cannot live without God. (*See* subtitle "Friendship" in the previous Chapter).

Atma Nivedana

Atma Nivedana is self-surrender to the Lord. Even an atom cannot move without God. He will take care of everything. Knowing this, the devotee offers himself entirely to God, including his body, mind and soul. He keeps nothing for himself. He has no personal existence, as he feels that he is a part of God. He considers himself as a mere instrument in the hands of God. It is God's duty to take care of him. He puts all the responsibility on God's shoulders.

Such self-surrender destroys egoism, sensual craving, anxiety, fear and worry. The devotee who has offered himself to the Lord has no desires, not even the desire for salvation. He is satisfied with the Love of God. There is nothing else to be attained.

Self-surrender stems from pure and absolute love for God. Nothing else can lead to total self-surrender. Hence *Atma Nivedana* is the highest expression of Divine Love. The devotee is totally one with God.

Complete self-surrender is itself attainment of God-consciousness. The mind merges into the pure Self. The individual soul merges into the Supreme Being. Man becomes God. The mortal becomes immortal. (For further details *see* Chapter 25. Cultivation of Bhakti: Positive Practices IV.)

Closing Remarks

The nine modes are spiritual practices through which devotees can attain God-consciousness. A devotee can take up any one mode or a combination of modes and attain the highest state.

In these nine modes one can find a synthesis of the three main Yogas – Karma Yoga, Bhakti Yoga and Jnana Yoga. *Padasevana* and *Dasya* are the practices of Karma Yoga. God is seated in all hearts. So serving God's feet means serving all beings. *Dasya* is regarding oneself as the servant of the Lord. The Lord is in everyone. So one should cheerfully serve all. This is Karma Yoga.

Kirtan, Archana and *Sakhya* are limbs of Bhakti Yoga. *Sravana, Smarana* and *Atma Nivedana* constitute the three-fold process of Jnana Yoga practice. In the modern age the Yoga of synthesis alone is the best to adopt for attaining rapid spiritual progress.

18
Obstacles and Remedies

Introduction

The spiritual practice is a lifelong process. Every minute, every hour, every day is an onward march. But the path is not a smooth one. It is rugged and full of hurdles. Obstacles are innumerable in this voyage. They act as stumbling blocks. If you overcome one obstacle, another is ready to manifest. If you control the sense of taste, another sense is simply waiting to assault you with redoubled force and vigor. If you remove greed, lust is waiting to hurl you down. If you drive egoism through one door, it enters another door. An intelligent understanding of the various obstacles in the Path of Devotion is absolutely necessary. Then alone acn an aspirant chalk out an appropriate remedial plan of action.

Sri Sankaracharya in his *Aparokshanubhuti* (Self-realization) lists the obstacles that arise during the spiritual practice:

> While practicing Yoga, there appear many obstacles such as lack of inquiry, idleness, desire for sensual pleasure, sleep, dullness, distraction, tasting of joy and the sense of blankness. One desiring the knowledge of Brahman should slowly get rid of such innumerable obstacles. (Verses 127-28)
> Patanjali also gives a similar list in his *Yoga Sutras*:
> Disease, dullness, doubt, lack of enthusiasm, laziness, lack of dispassion, False perception, lack of concentration, and falling away from the state reached are the distracting obstacles. (1. 30)

Most of these obstacles are created by our own careless ways of life. They include neglect of moral discipline, over-eating, over-sleeping, meaningless work, talking too much and such other things. Irresponsible ways of life are incompatible with spiritual aspirations. What we are today is the result of our past actions. Yet we are not just at the mercy of our past actions. What we are doing right now is, in the words of Vasistha, the *Guru* of prince Rama, "infinitely more potent than the past." Vasistha further says, "Fate is nothing but the self effort of a past incarnation... But there is no power greater than right action in the present......Hence ....one should overcome evil by good, and fate by present effort." (*Yoga Vaisistha*).

The obstacles may be grouped under external and internal or mental problems. They are, however, not mutually exclusive, but inter-related. The external obstacles include poor health, improper dietary habits, distractions, and evil company. The internal obstacles include ignorance, egoism, attachments to sense pleasures, lust and greed, laziness, doubt, delusion, depression, vanity and pride, conceit and such others.

If we carefully look at these obstacles, we would see that there are no real external obstacles. The so-called external obstacles are only situations arising out of internal problems. So the real obstacles are those that exist within us, the impure tendencies that hold us back from experiencing God's love. These tendencies consist of egoism, desire, selfishness, attachment, sloth, carelessness, delusion, doubt and so on. These tendencies manifest in various situations. And one by one they have to be eliminated by strengthening our pure tendencies. We have to continue our spiritual practice, until finally the great obstacle – the root cause of all others – our ignorance of our real nature gives way to Self-knowledge.

The major obstacles to be overcome and the remedial measures to be taken up are described below.

Ill-health

Spiritual practice is not possible without good health. A sickly body is not fit for the practice. Sickness causes aches and pains, and also affects the mind. It causes dullness and laziness. Therefore, an aspirant should try to maintain good health by regular exercise, asanas, Pranayama, proper dietary habits, moderation in diet, work, recreation, sleep, etc. He should avoid taking drugs as far as possible. He must take recourse to natural cure, and always keep a cheerful attitude of mind. Cheerfulness is a powerful mental tonic. There is intimate connection between body and mind. If one is cheerful and mentally peaceful, the body is also healthy.

Impure Food and Improper Dietary Habits

The mind is formed out of the subtlest portion of food. So if food is impure, the mind also becomes impure. Meat, fish, eggs, stale unwholesome food, onions, garlic, etc. should be avoided, as they excite passion and anger. Liquors and narcotics should be strictly abandoned. Chilies, condiments and spiced dishes must be rejected. Fat-free vegetarian diet is the most appropriate food for maintaining good health.

Dietary habits also influence health. Untimely eating, eating more frequently, over-eating and eating heavy items like meat cause digestive problems. Most of the diseases are caused by over-eating and over indulgence. Moderation in eating, working, recreation and sleeping is the key to sound health. If there is no hunger, you must not take any food. The night meals should be very light. Fasting for a day per week is desirable.

Ignorance

Ignorance is the root cause of all obstacles. In the worldly sense, ignorance refers to the lack of knowledge pertaining to the mode of living, health, diet, work and rest. It leads to ill health and wrong habits.

In the spiritual sense, ignorance is wrong understanding of our real nature. It means considering the impermanent as permanent, the impure as pure, and the non-self as the Self. (*Yoga Sutras*, 2. 5). The body, senses and mind are impermanent and non-self. Spirit or Self or Atman is eternal, real, never changing and omniscient.

Pure eternal joy and peace are our real nature. Our ignorance veils our insight. We fail to perceive our inner eternal peace and joy and instead seek it in the external world. We cling to impermanent worldly pleasures and fall into the whirlpool of endless desires and the consequent sufferings and bondage. When we attain enlightenment through the practice of Yoga, then only ignorance disappears.

Egoism

Egoism is the feeling of "I, me and mine." This prevents us from realizing our true nature. It is the ego that is the basis of all mental dramas—desires, attachments, aversion, selfishness, karmic bondage and so on. It creates all kinds of problems, troubles, anxieties and fear. It disturbs our mental peace. Ignorance of our real nature is the true cause of egoism. Ego veils our eyes, and ignorance results. In reality we are Spirit. We have body, senses and mind, but when we forget that we are Spirit and identify ourselves with the body, senses and mind, the sense of ego intervenes, and we forget our super-conscious nature.

It is very difficult to get rid of the ego. The mind's nature being what it is, the ego perpetually reasserts itself. Continuous sincere effort is, therefore, essential. By self-analysis, deep Devotion to God, and total self-surrender to God we overcome our ego. We must read the life stories of great sages and saints and learn lessons from their lives. We are nothing when compared to those great souls. Let us realize that we are mere instruments in the hands of God. It is He who functions through us. Our minds, intellect, bodies and senses all function only because of the consciousness within. That life force is part of the Cosmic Consciousness, God. We should realize this Truth.

Delusion

Delusion means seeing the unreal as real. The most basic delusion is taking the body-mind complex as our real "I." This basic delusion is the premise upon which all other delusions are based. Because of this delusion we create bondage to a false world, and impose on ourselves unreal restrictions and limitations.

When our identity is rooted in Consciousness, there is no problem with delusion. In true knowledge there is no identification with the body, no thought of being anything. It is when we think we are something that delusion creeps in. To overcome delusion we need to recognize the truth that we are Consciousness. One who lives in the awareness of this Truth is free from delusion. If we know exactly how we were deluded, the delusion would leave us. Since we have no way of knowing all the ways in which delusion arises in our life, we need the Guru, one who lives in the awareness of the Truth.

Doubt

An aspirant begins to doubt whether God exists or not, whether he will succeed in realizing God or not. Lack of faith is a dangerous obstacle in the spiritual path. The aspirant slackens his efforts when these doubts crop up. He should clear his doubt immediately by discussing with his Guru or any illuminated saint. He should also listen to spiritual discourses and reflect over the spiritual messages.

One should not get dejected after some practice. Even a little practice has its own effect. You may not be able to perceive it. You must have unshakable conviction in the existence of God and in the efficacy of spiritual practice. You should read the stories of realized sages and derive inspiration and confidence from their examples. You must develop the virtues of patience, perseverance and determination. Make an affirmation, "I will realize God right now in this very birth."

Laziness

Laziness is a *tamasic* quality, an innate tendency toward lethargy. It is a great obstacle to spiritual progress. When there is sloth, there is little inclination for work. The source of sloth is often purely physical. Due to ill health or poor diet, lack of exercise and irregular habits, the body becomes weak and toxic. The weakness produces a fogginess in the mind. It results in disinclination to engage oneself in devotional practices. You must develop the habit of performing your duties punctually and with great enthusiasm. The great clue to conquering laziness is hard work, effort and willingness to arouse you. It is the initial inertia that must be overcome.

Sensual Desires and Pleasures

Craving for worldly things and sensual pleasures is a great obstacle in the spiritual path. The mind that goes after material things and pleasures through sensual desires cannot concentrate on God. The mind becomes restless like a mad monkey. There is no end to desires. They lead to anxieties, worries and disappointment.

Hidden desires for fame, name, power, wealth and sex, and subtle memories of pleasures enjoyed earlier may cause serious distractions in the mind. Every little unsatisfied desire and every indecisive thought will distract our concentration. Success in Yoga will not come until we disband the conflicting hosts of desires that perpetually carry on their civil war within us.

We must contemplate on the fleeting nature of worldly pleasures, and discipline our mind and senses. Study of scriptures and company of wise people will help us to develop dispassion.

But the mind often refuses to completely give up the craving for sense-objects and sensual pleasures. Through the force of spiritual practice, the desires may get suppressed for sometime. Then all of a sudden, the memory of past experience surfaces, causing mental disturbance and loss of discrimination. The mind is excited to move again outward to sensual objects. In the *Bhagavad Gita* Lord Krishna points out this danger:

> "O son of Kunti, the excited senses impetuously carry away the mind of even A wise person, though he be striving (to control them). The mind that yields to The wandering senses carries away his discrimination as the wind carries away A boat on the waters." (2. 60, 67)

Just as the wind carries away a boat from its course, so also the mind attracted by sensual objects carries away the seeker from his spiritual path.

The seeker, through intense practice of dispassion and austerities, may succeed in controlling the senses and abandoning sensual pleasures, but the relish or taste for them may still remain. So he must be alert and careful. He should be ever watching the mind vigilantly. He must nip the taste for pleasures in the bud by practising severe austerities.

Lust and Greed

The greatest obstacle to spiritual practice is 'lust and greed.' In the words of Sri Ramakrishna, a human being can realize God if he wants to, but he madly craves the enjoyment of sex and worldliness. For sex one becomes the slave of another, and so loses one's freedom. Attachment to sex diverts one from the path leading to God. 'Lust and greed' is the cause of bondage. It is the root of the tree and desires for power, position, wealth, etc. are its branches and twigs. One cannot completely get rid of the six passions: lust, anger, greed and the like. Therefore, one should direct them towards God.

One who has renounced the pleasure of the spouse has verily renounced the pleasure of the world. God is very near to such a person. The 'lust and greed' is the only cloud that hides the Sun of Knowledge. The darkness of the mind is destroyed only when a person stands apart from 'lust and greed,' and practices austerity and spiritual discipline. Only then does the cloud of his ego and ignorance vanish. Only then does he attain the Knowledge of God. 'Lust and greed' cannot do any harm to the person who lives in the world after attaining God. Only then can he lead a detached life in the world as King Janaka did. But he must be careful at the beginning. He must practice strict discipline in strict solitude.

In the case of householders, after the birth of one or two children, husband and wife should live as brother and sister and talk only of God, and help one another on the path of spirituality. None can taste Divine bliss without giving up his / her animal feeling. A devotee should pray to God to help him/her to get rid of this feeling. It must be a sincere prayer. God will certainly listen to our prayer if it is sincere. (Source: *The Gospel of Sri Ramakrishna*)

Free Mixing with the Opposite Sex

Even if a spiritual aspirant renders selfless service without attachment to the fruits thereof, it is not safe for him to engage himself in those activities that would compel him to freely mix with the opposite sex or atheists or entangle him in the meshes of worldly riches. In the early stages of spiritual practice contact with such people is so dangerous that Narada admonishes a strict warning: not even to hear or read about them. (*N.B.S.*, 63). Even through hearing stories about them, an aspiring devotee may become interested in those things and be gradually tempted to give up the service of the Lord and run after worldly pursuits. It is therefore necessary for a modern seeker of Divine Love to be careful to keep away from novels, pictures and stories woven round sexual passion and from the life and work of atheists and materialists. The insistence of Narada on this point is restated in a summary way by Sri Ramakrishna in the nineteenth century in his well-known phrase 'lust and greed' (*kamini-kanchan*). Hence spiritual aspirants should take sufficient precautions against sexual temptation, self-aggrandizement and lapse into atheism. Even if a devotee is above all temptations, it is better on his part to observe these restrictions in order to set an example to the world.

Depression

An aspirant may get moods of depression sometimes. These moods may be due to indigestion, cloudy conditions, and revival of old tendencies from within and influence of bad company. The cause should be removed. Depression should not be allowed to overpower. It must be removed immediately by cheerful thoughts, a brisk walk, singing, prayer, Pranayama and reading some elevating portion of a scripture.

Distraction

Scents, soft bed, novel reading, dramas, movies, vulgar music, dancing, flowers, company of opposite sex, fat and spicy diet – all these excite passions and cause disturbance of the mind. Too much salt, too much chilies and too much sweets cause intense thirst and disturb spiritual practice. Too much talking and too much exercise also disturb the mind.

Impulses disturb spiritual practice. All obscure sub-conscious impulses should be controlled by discrimination and will. Sex-impulse and ambition are real distracting factors. They carry on guerilla warfare. They attack the devotee again and again. They appear to be thinned out for sometime. They get revived often. They should be extirpated by great efforts, self-inquiry, discrimination and sublimation into intense devotion to God.

Conceit

Even a devotee of God has his own conceit that he is a greater devotee than others are. Sri Ramakrishna taught us to say before God, "I am nothing. You are everything." Even when you say to God, "I am your devotee, your child, your servant" there is some trace of pride. There is no conceit at all when you say, "I am nothing." One who is intoxicated always with the thought of God is the one who is free from the ego-sense, not others. When you surrender yourself to God, your Ego will die and the desires will disappear. When the tree is cut down, the birds on the tree fly away.

Carelessness

When we really care about something, we usually do it well. The interest in spiritual practice is not something we possess full-blown at the start. It must be developed over a period of time, giving undivided attention to it. We must be clear about our goal. We have to ask ourselves – whether what we are doing will lead us toward the goal. The key to being careful is contemplation. We have to ponder over our practice with great care to weigh it and reflect on it.

We have to look at the subtle attitude with which we do our practice. We must do it consciously and with an awareness of its relevance to the ultimate goal.

Vanity and Pride

A devotee of the Lord is likely to fall a prey to feelings of vanity and pride when he is engaged in good altruistic work as a result of his own estimation of his achievements as great. He is warned against this danger. He should curb such feelings by nipping them in bud. (*N.B.S.*, 64). Even when he helps the needy, he should consider that as a service to the Lord, and should therefore be grateful to the recipient for having given him an opportunity to have the joy of service. If the devotee cultivates this attitude, he can easily avoid pride and vanity, which are usually found in ordinary philanthropic activities.

It is very difficult to curb completely selfish instincts and feelings all at once. Forcible repression is not advisable as it would lead to injurious results. Therefore gradual sublimation of those instincts is necessary. A safer course is to make use of these passions in the practice of Devotion to God. Anger may be directed toward the obstacles to Bhakti; it will then take the shape of renunciation and dispassion. Pride may be entertained in association with the feeling that he is a child or servant of the Lord. Even this pride is not laudable. However, this type of pride is not bad as worldly pride, for it would gradually wear out as devotion reaches its perfection. In this way he may sublimate all the passions that normally arise in the course of a person's conduct in society.

Indulgence in Arguments

Indulgence in arguments and controversial discussions with others are an obstacle to spiritual practice. It causes distraction and disturbance to the mind. By arguments you get nowhere. Arguments cannot conclusively establish the existence of God. By arguments neither an atheist can prove the non-existence of God, nor can a saint prove the existence of God. God is not something that can be shown. God has to be directly experienced within. "The Self is not known," says *Katha Upanishad*, "through the study of the scriptures, nor through subtlety of the intellect, nor through much hearing. But by him who longs for Him is He known. Verily unto him does the Self reveal His true nature." (1. 2. 23). So a true devotee should avoid vain disputations and arguments.

Evil Company

The effects of (an) evil company are highly disastrous. An aspirant should avoid the company of those who speak lies, who commit adultery, theft, cheating and under-hand dealing, who indulge in idle talk, back-biting and tale-bearing, who are greedy, and who have no faith in God and in moral values. Evil company pollutes the mind. The little faith in God and moral life also disappears. "A man is known by the company he keeps." This is a wise maxim. Just as a young plant is to be well fenced in the beginning for protection against cows, etc., so also an aspirant should protect himself against all sorts of evil influences. (*See also* chapter 21. Cultivation of Bhakti: Negative Practices II.)

Break in Practice

Some aspirants leave practice after a while. They expect quick results and psychic powers. When they do not get some they give up practice. Sometimes the aspirant is sidetracked. He loses his way and walks in some other direction. He misses the goal. Sometimes he gets false contentment. He thinks he has reached his goal and stops all practices. Sometimes he is assailed by temptations. Sometimes he is careless, lazy and indolent, and he cannot do any practice. Therefore an aspirant should be eternally vigilant like the captain of a ship.

Breaks in practice not only nullify the good effects of previous practice, but also often cause permanent injury. Breaks become injurious if they are caused by wanton negligence or temptations of the senses. Any conscious yielding to such temptations makes a devotee weaker. Such yielding has to be guarded against.

Sustaining interest is essential for avoiding negligent breaks. A good way of keeping up interest is to provide sufficient variety. Worship, *Japa*, chanting, study of scriptures, listening to spiritual discourses, service and other modes of devotion may be given their rightful share in the scheme of practice. That it is possible to have these varieties is illustrated by Sri Ramakrishna in his own life and teachings.

Religious One-sidedness

Religious one-sidedness is a major obstacle to real spiritual awakening. Narrow loyalty to a particular faith and condemning or rejecting other paths as untrue would never help developing the cosmic vision of the Truth. No faith is superior or inferior to any other faith. The scriptures say: the Truth is one, the paths are many. That is, the various faiths are just different paths leading to the same goal of God-realization Sri Ramakrishna's experience is an illustrious proof for the unity of all religions, as described in chapter 6. The various religions have come into existence because of variations in people's temperaments, tastes, outlook and environment. God is everything. Everything is His manifestation. An aspirant should develop a universal religious outlook. Then only the inner power of illumination can fully blossom.

Conclusion

The spiritual path may, in the beginning, appear to be very hard. Renunciation of sensual objects gives pain at the outset. But if you make a strong determination and firm resolve, then it becomes very easy. You develop interest and experience joy in your spiritual practice. Your heart expands. Your outlook becomes broadened. You get a new, wide vision. You feel the help from the invisible hands of the Indweller of your heart.

Your doubts are cleared by themselves by getting answers from within. You can hear the sweet voice of God. You experience the indescribable thrill of Divine ecstasy from within, everlasting joy and peace. This gives you new strength. The footing on the path becomes firmer and firmer. You are backed up at all times by a mighty Cosmic power that works everywhere.

Following the advice of Sri Swami Sivananda, assimilate the soul-awakening with spiritual thoughts or divine ideas. Saturate the mind with thoughts of God, Divine glory and Divine Presence. Then only you will always be established in the Divine Consciousness and attain Realization. Always remember the triplet: "Assimilation, Saturation, Realization."

19

Cultivation of Bhakti: Introduction

The natural perfection of the human soul is manifested, only when the various faculties of the mind are purified and coordinated harmoniously. Spiritual practices are meant to effect this through the cultivation of the various faculties of the mind. Bhakti yoga is mainly concerned with the purification of the emotions. The emotions are cultivated through Devotion to God. It is a means to attain Supreme Devotion. There are various disciplines that help to attain this Devotion. The disciplines fall into two groups: Positive and Negative. Negative Practices include overcoming *Maya*, renunciation of the ego, transcending *Gunas,* non-attachment, giving up fruits of action, avoiding vain discussions, shunning evil company and transcending devotion with motives.

The Positive Practices consist of preparatory discipline, cultivation of virtues, seeking holy company, listening and singing, surrender to God, worship, prayer, *Japa*, service, dedication of actions and reflecting upon the scriptures. All the practices are actual experiences of great teachers. They constitute the essence of Bhakti discipline. They are universal practices and may be adopted by all spiritual seekers irrespective of birth, faith and sect.

An aspirant may begin with either Negative or Positive Practices, depending on his capacity, opportunity and convenience. But real success comes only when both the Positive and Negative Practices are pursued simultaneously. Negative Practices prevent retrograding and Positive Practices lead to progress.

Sow the Seed of Devotion in Youth

True Devotion to God requires various qualities such as a loving heart, compassion, contentment, purity, sense of equality, balanced mind and sincerity as explained in tchapter 11. Qualifications and Qualities of a Devotee. None of us possess those qualities to the fullest extent. We have to cultivate them. It takes a great deal of time and much patient, painstaking effort to cultivate even a little of these qualities. We will have to utilize every available minute in trying to cultivate Devotion to God in our heart. Therefore, the best time to start is now, when the body and mind are young and can easily be directed toward God.

Devotion is nothing to do with age, caste, position, rank or gender. Generally people say, "I will do my devotion practices when I retire." This is a serious mistake. How can you perform intense spiritual practices after all your energy has been exhausted? You will not be able to sit for even fifteen minutes in a steady posture. You will not have the strength to discipline the senses. The spiritual seed of devotion must be sown in your heart when you are young, when your heart is untainted and pure. Then it will strike deep root, blossom and bear fruit in your old age, when you retire. Then you will have no fear of the Lord of Death.

When the body dies the soul takes on a new body. It is very rare to be born as a human being. It is only in human form that we can strive to realize God and attain eternal bliss, and not through any other form. Rarer still is to have a desire to follow the spiritual path, having attained a human body. Rarest of all is to have intense thirst to realize God. This gift is due to the Grace of God. Having obtained a human birth, you would be foolish not to strive for God. This is suicidal. You are destined to achieve the ultimate goal of life. Why not begin now? Learn to love God as much as you love your own self? With real understanding and faith give up the desire for fleeting pleasures from the transient objects of this world, and fix your mind on the feet of the Lord.

"Desires are never quenched by enjoyment. Therefore even the slightest desire must be nipped in the bud by abstinence, as it would otherwise lead to perdition.. Hook the fish of desire by abstinence." (*Yogavaisitha*, IV, 77. 81,83).

"Renounce all attachments mentally. Sublimate them into intense Devotion to God. When the mind turns away from sense-objects, the senses too will follow. The total renunciation gives freedom from misery."

(*N.B.S.*, 35)

No Bargaining with God

If you have tasted pure milk, will you run after water mixed with some white powder? Even so, if you have once tasted the sweetness of Divine Love, what else can you desire but God alone! The real devotee says, "I do not want anything from my Lord. Let my mind be ever fixed at His Lotus Feet. Let my soul ever cling to Him."

The boy Prahlada was an ideal devotee. Pleased with his devotion, the Lord offered him anything that he desired. Prahlada replied, "My Lord, kindly do not tempt me by offering me boons. I have come to you to get rid of desires, not to satisfy them. Grant me pure love for Thee. I seek shelter in Thee."

In the name of devotion we should not do shop keeping with God. He who asks for a boon is trading, bargaining or doing business with God. He becomes a trader in love, a trader in religion.

A college student of India bargains with God and prays, "O Lord Ganesha! Kindly help me pass in my examination this year, I shall offer unto Thee a hundred and eight coconuts." He who bargains with the Lord in this way cannot be a true devotee of God.

Suffering and Sorrows

So God is not to be approached in the above manner. In the higher stages of devotion, the devotee even refrains from asking God to take away his miseries and sorrows, his pains and diseases, and his trials and difficulties, because he knows that they all come from Him and are His gifts. A coward cannot tread the Path of Devotion. It calls for heroism of the highest kind.

Always remember the prayer of Kunti Devi, the mother of the Pandavas. She prayed to Lord Krishna, "O Lord! Let me always have adversity so that my mind may be ever fixed at Thy Lotus Feet." During suffering we remember God. Suffering is a great eye-opener. It turns our mind away from petty concerns in the world and directs it towards God. Had there been no suffering in this world, Man would never have tried to free himself from the bondage to this world. He would remain fully satisfied with the trifling pleasures of the world. Sufferings and difficulties are therefore blessings in disguise. When we face difficulties we are strengthened. A sportsman's true color is seen only when he confronts a strong opponent. Similarly our Devotion to God is seen when we are in distress. Distress makes us to intensify our Devotion to God.

God puts His devotees through severe tests and rigorous trials before finally blessing them with His *Darshan* (Vision). He at the same time also gives them the necessary power of endurance, spiritual strength and patience to overcome these tests. We must have total trust and faith in Him.

Indeed God takes away the sufferings of His devotees. What this really means is that the true devotee no longer sees the sufferings and difficulties as being such. The same experiences that would leave an ordinary person helpless are seen by the devotee as blessing from God! The difficulties are there, but they are no longer a burden to the devotee. This is how God shows His Divine mercy to His sincere devotees.

Does this mean that it is really useless to pray for worldly things, for success in some undertaking or for relief from sufferings and sorrows? Do such prayers have no benefit? Such prayers also have their benefit. Even though they are somewhat selfish, eventually they will lead us to true devotion. If God grants us our worldly prayers, we will develop faith in Him and get closer to Him. Ultimately, our devotion will become pure and unselfish. The devotion of most of us starts in this manner. So such prayers to God should not be frowned upon. Such selfish prayers have changed the course of many devotees and made them great devotees. God's ways are indeed mysterious. If He wishes to take us to His Lotus feet, He may use any approach.

20

Cultivation of Bhakti: Negative Practices-I

How Bhakti is to be cultivated through certain disciplines was introduced in the previous Chapter. The disciplines that help to cultivate devotion fall into two groups: Positive and Negative Practices. The Negative Practices prevent regressing, and Positive Practices lead to progress. The Negative Practices are taken up for discussion. with the Negative Practices of 'Overcoming Maya' and 'Renunciation of Ego' dealt with first in this Chapter.

OVERCOMING MAYA (COSMIC ILLUSION)

What is Maya?

Man is a paradoxical being. At the one end, he is simply an ordinary animal, but at the other end, he is Divine and a blissful Spirit or Atman. Divinity is latent in man. But he generally not aware of it. Because of Divinity in him, he should be ever contented, but he longs for the enjoyment of the senses and feels miserable when he does not get sense enjoyment or loses it. The Divinity in him should make him feel ever peaceful, but often he is restless and agitated because of anxieties and disappointments. There is thus a glaring mystery about man. He is ignorant of his true nature. This ignorance of Truth is called Maya.

Man is far above other animals in evolution, but this is not fully recognized. Even a learned person does not behave better than an animal. A cow, for example, is naturally attracted towards a person holding a bunch of grass for it; but it runs away from him if he goes near it, shouting and holding a big stick in his hand. Similarly, even a learned person instinctively runs after sense pleasures and tries to run away from danger and pain. He identifies himself only with his body-mind complex. This identification with the body is the root of his ignorance, and makes him run after pleasure and flee from pain. He is in the bonds of Maya. Maya can be transcended by Yoga, as shown by Sages and Saints since ages.

The Self (Atman) within us is one with Brahman, the Supreme Consciousness that we call God. That Self is the unchanging Reality within us. But through ignorance we wrongly identify the Self with the non-self, i.e., the body, mind and senses. Thus we become unaware of our Divinity, and consider ourselves finite, limited and bound. We are unable to turn away from the world of sense-objects. Because of our attachments to them, our soul is subject to the ups and downs of worldly life, and the cycle of births and deaths through repeated embodiments.

Crossing the Maya

Who does cross the ocean of Maya? One, who gives up all attachments, who serves a spiritually elevated person (or Guru) and who becomes free from the sense of "I and mine," crosses Maya. (*N.B.S.*, 46).

The worldly life is compared to an ocean. Pleasure and pain, hunger and thirst, birth and death are waves; lust, greed, anger, hatred, passion and other impurities are the terrible aquatic creatures that endanger the swimmer who dares to cross it; repeated embodiments are the numerous whirlpools; attachments and infatuations are the undercurrents; and the effects of the past bad actions (*Karmas*) are the sharp rocks hidden beneath the surface. Thus this ocean of worldliness surges with sorrow and suffering. While an aspirant following the Path of Wisdom relies on self-effort like one that tries to swim across the ocean with his bare arms, a devotee secures the comfortable ship of Divine Grace and is safely taken to the other shore by the Lord who captains the ship. In the *Bhagavad Gita*, Lord Krishna says,

"How hard to break through is this, My Maya made of the Gunas!
But he who takes refuge in Me only shall pass beyond Maya;
He, and no other." (7. 14)

Requirements

How does a devotee become fit to receive the Grace of the Lord in order to cross the 'ocean' of Maya?

First, a devotee has to renounce all attachments by the cultivation of discrimination between real and unreal. (*N.B.S.*, 35). Attachments are of two types: external and internal. External attachment means attachment to sense-objects of the world. This is curtailed by adopting the ideal of simple living. It involves the practice of *asteya* (non-stealing), *aparigraha* (non-covetousness or not possessing objects beyond one's needs), and *uparati* (withdrawal from objects of sensual pleasure).

Internal attachment consists of subtle desires (*vasanas*) of unconscious mind. One who practices physical renunciation of objects, but continue to cherish the sense-enjoyment in the mind cannot achieve much progress on the spiritual path. Lord Krishna says in the *Bhagavad Gita*,

"He who, restraining the organs of action, sits thinking of the sense-objects
In mind, is of deluded understanding and a hypocrite." (3. 6)

The attachments are thinned as the devotee's mind begins to enjoy the sweetness of Divine Love. They are finally eradicated when devotion matures into *Para Bhakti* (Supreme Devotion).

Second, the devotee has to render service to his Guru. "He who serves the Guru attains the Knowledge of Brahman, just as only he who digs with a spade gets water," says *Manusmrti* (2. 218). By serving the Guru wholeheartedly, the devotee manifests his humility and devotion to the Guru. Then, through the Grace of the Guru and by following his teachings, the cloud of ignorance disappears, and the Knowledge of God dawns.

Third, the devotee has to become free from the sense of possession. He achieves this by practising discrimination and realizing that the worldly objects are ephemeral and unreal. When he purifies his heart through spiritual discipline and devotion, he becomes increasingly aware that nothing belongs to him. When he loves the highest, everything low naturally falls away. The ancient Tamil saint Tiruvalluvar in his *Tirukural* says,

"Cling to the One who clings to nothing;
And so clinging, cease to cling." (350)

Fourth, freeing himself of all distractions, the devotee has to sit in solitude and intensify his longing for God. (*N.B.S.*, 47). Deep Meditation or *Japa* cannot be practiced in a noisy or clamorous place. Therefore he must set aside some time every day when he can be alone and undisturbed. The scriptures state that early morning between four and six o'clock and evening during sunset are best suited for the practice of Meditation and *Japa*.

In an advanced stage, the devotee discovers that the Self is ever alone, independent of all relationships, free of desires, without localization in the world of time and space, and without home or property. When this stage is reached, no matter where he is, he will be free of all distractions and constraints and be in solitude at all times.

Fifth, in order to cross the Maya, the devotee should perform all actions without selfish motives and dedicate them to God and thus free himself from the bondage of the world. (*N.B.S.*, 47). How does bondage arise? Performance of actions with selfish motives and attachment to the fruits of actions cause bondage and subjection to the cycle of births and deaths. This is the effect of the Law of Karma. Every action produces a corresponding effect. It is our selfish motives and our attachments to actions and their fruits that bind us. So the devotee should perform actions without selfish motives and dedicate them to the Lord.

Sixth, the devotee has to become free from the three Gunas: *Sattva*, *Rajas* and *Tamas*. (For details *see* chapter 21. Cultivation of Bhakti: Negative Practices II.). *Tamas* hurls one down to lethargy and delusion; *Rajas* makes one overactive and passionate. So the devotee has to cultivate *Sattvic* mode of mind through intensification of devotion and purification of the mind. With the predominance of *Sattva* one can overcome *Rajas* and *Tamas*. He then attains intuitive vision of the Self and rises beyond the three Gunas. "When the milk of his mind has been turned into butter of Divine Love," the devotee is freed from the bondage of the world and the effects of the Gunas.

Seventh, a devotee has to renounce his concern for acquiring worldly objects and for preserving them. Most people are ever pre-occupied with this two-fold concern, and this mental involvement is a great obstacle on the Path of Devotion.

A devotee who totally surrenders himself to God need not have to care even for his personal needs. God takes care of his living. (*N.B.S.*, 47). Therefore his mind becomes free and established in God. In the *Bhagavad Gita*, Lord Krishna says,

> "To those persons who worship Me alone, ever united, thinking of no other,
> I secure what they need and protect what they possess." (9. 22)

Thus freed from all worries and preoccupations, the devotee is able to develop one-pointed devotion to the Lord.

Eighth, A devotee has to give up the fruits of actions, and renounce the actions themselves. (*N.B.S.*, 48). Actions performed with selfish expectations are the root-cause of repeated embodiments. Therefore, a devotee must pursue the path of selfless action (*Nishkamya Karma*), and perform actions without selfish expectations. When he dedicates the actions and their fruits to the Lord, he renounces them. Then he is no longer their doer and thus becomes free from their binding effect. The statement that a devotee renounces all actions does not mean that he becomes absolutely inactive. Such a state is impossible. As Lord Krishna says in *Bhagavad Gita*,

"Verily none can ever remain for even a moment without performing action;
For, everyone is made to act helplessly by the Gunas born of Nature."
(3.5)

Ninth, a devotee has to go beyond the pairs of opposites. (*N.B.S.*, 48). Pleasure and pain, praise and censure, gain and loss and such others are pairs of opposites. They cause agitation and distraction to the egoistic human mind. But a devotee whose mind is purified by his spiritual disciplines and by his deep Devotion to God maintains a balanced mind. He is not elated due to delightful events, nor is depressed by painful events. Immune to the pairs of opposites, he enjoys the indescribable sweetness of Divine Love at all times.

Tenth, in order to cross the Maya, a devotee has to renounce even the rituals and ceremonies prescribed by the Scriptures, and attain uninterrupted flow of Divine Love. (*N.B.S.*, 49). The ritualistic actions are performed for a heavenly reward. Heaven is not the ultimate goal. The performer of meritorious actions may enjoy heavenly pleasure, but after the effects of those actions are exhausted, he has to come back to the earth. But the devotee who attains Supreme Devotion realizes God and thus attains the final Goal of life. He has no more birth. Even if the rituals serve as a means, they may be undertaken in earlier stages, just as a boat is used to take us across a river; but when we reach the opposite bank we have to get down and walk away leaving the boat. In the same way the devotee immersed in the nectar of Divine Love transcends the scriptural commands.

Summing up, the devotee who fulfills the above requirements – renouncing all attachments, rendering service to the Guru, freeing himself from the sense of possession, taking recourse to a solitary place, rooting out the bondage of the world, transcending the three Gunas, renouncing the concern for acquiring and preserving worldly objects, giving up the fruits of actions, renouncing all actions, going beyond the pairs of opposites, transcending scriptural sanctions and attaining Supreme Divine Love – crosses the Maya.

These requirements are similar to the method of God-realization described by the Lord Krishna in the following verses of *Bhagavad Gita*:

> "Endowed with a pure intellect, controlling the self by firmness,
>   Relinquishing sense-objects and abandoning both aversion and attraction,
>   Dwelling in solitude, eating but little, with speech, body and mind subdued,
>   Always engaged in meditation and concentration, taking refuge in dispassion,
>   Having abandoned egoism, strength, arrogance, desire, anger and covetousness,
>   Free from the notion of 'mine,' and remaining peaceful –
>   He is fit for becoming Brahman.
>   Becoming Brahman, serene in the Self, he neither grieves nor desires;
>   The same to all beings, he obtains Supreme Devotion unto Me." (18. 51-54)

Having crossed the Maya, the perfect devotee helps others to cross it. (*N.B.S.*, 50). The realized devotee is not just satisfied with his own perfection. As an instrument of God he helps others to cross the Maya and attain perfection. Supreme Devotion is not a selfish end. Having attained this illuminating Love, the devotee becomes a true Guru to help others to reach the goal that he himself has reached. Thus the Incarnates (*Avatars*) such as Buddha, Jesus and Sri Ramakrishna, and the realized Sages and Saints since ages became the saviors of humanity.

RENUNCIATION OF EGO

Ego or the feeling of "I, me, mine" is the source of desire for sense objects and pleasures, selfishness, attachments to desirable things, aversion to unpleasant things, negative emotions such as lust, greed, anger, hatred, jealousy and others, and selfish actions. All these impurities vitiate the mind. The mind becomes restless and wandering. There is no peace, but there is ego. The ego makes it impossible to experience happiness in our own being, to experience the Truth.

Once there was a man who heard that he could attain supernatural powers by performing a certain practice. So he did that for years. Many years later he attained the power to multiply himself into thirty identical forms. He was thrilled. He was happy. He went from place to place performing this trick. Everyone appreciated his miracle. This boosted up his ego. He kept on doing this feat.

However great one is, one cannot avoid one thing in life, and that is death. The man's final day came. Because of his practices, he knew that the hour was approaching. When he heard the bells of the messenger of death, he multiplied himself into thirty identical forms. The messenger just stood there. He was supposed to take one person, but here were thirty persons who all looked the same. He was confused. He went back to the Lord of Death and reported, "Lord! there were thirty identical persons. What could I do?"

The Lord of Death knows the weakness of the human beings very well. He called another messenger and whispered into his ear, "Praise him! Praise him to death!"

This messenger went to the man, who again multiplied himself into thirty forms. The messenger began to praise him, saying, "Wonderful! You are a great man. You are magnificent. You are incredible. There is no body like you. You are the best. You have surpassed God in your power of creation!"

As the adjectives got more and more superlative, the balloon of the ego kept getting bigger and bigger, and all the thirty heads swelled up. The man was totally intoxicated with the praises of his feat.

Finally the messenger said, "But there is one tiny mistake." The real person jumped forward and shouted, "What's that?" The messenger of death put the rope around his neck and dragged him away.

We think that disease or old age is the cause of death, but it is the ego that is the cause of death. Death is fine if it is death in the sense of leaving this world, but you die over and over while you are living. The *Kena Upanishad* says,

"If here a person knows It (Brahman), there is Truth, and if here he knows It not, there is great loss." (2. 5)

This means that if you have not experienced the Truth, your life in this world is a waste, i.e., your life is nothing but death.

The ego prevents us from knowing what we really are. It does not allow us to turn within. We must make determined effort.

Therefore an aspirant of God-realization has to renounce his ego and its offshoots of lust, greed, desires for sense-objects and attachment to the fruits of actions for the purification of the mind and heart. Only when the heart is pure, he can develop pure love and attain God.

Attitudes

In order to renounce egoistic involvement in life, the following attitudes are to be developed:

1. The Attitude of worshipping God (*Puja Bhava*). A devotee must feel that whatever he does is worship or expression of Devotion to God.

2. The Attitude of Detachment (*Anasakti Bhava*). A devotee should remain mentally detached from actions performed and from their fruits. He should reflect upon the well-known statement of Lord Krishna in the Gita: "Your right is to perform duty alone, and not to the fruits thereof." (2. 47). Everything happens according to a Divine Plan. So a devotee should do his duty without any expectation. No expectation, no disappointment.

3. The Attitude of being God's Instrument (*Nimitta Bhava*). The devotee should consider himself as an instrument in the Hands of God. With this attitude he recognizes the operation of Divine Will through him. He allows himself to be led by Divine Will.

4. Witnessing Attitude (*Sakshi Bhava*). As the mind is purified, the devotee develops a subtle attitude of being a witness to all the favorable and adverse developments in life. It is the mind and senses that operate driven by the forces of the *Gunas*, while the Self in him is always a detached witness or spectator.

5. The Attitude of Surrender to God (*Ishwararpana Bhava*).
   With the increasing purity of heart, a devotee surrenders himself
   to the Divine Self. When the process of surrender is complete,
   the devotee attains the goal of life.

Natural Process

What is the nature of renunciation of a devotee? His renunciation is
the most natural. It is easy, smooth-flowing as the things around us.
We see the manifestation of this sort of renunciation every day around
us. A person loves his own city; then he begins to love his country,
and the intense love for his little city drops off smoothly, naturally.
Again he learns to love the whole world; then his love for his country,
his intense fanatical patriotism, drops off without hurting him.

In human society, the nearer a person is to the animal, the stronger
is his pleasure in the senses; and the higher and the more cultured a
person is, the greater is his pleasure in intellectual and other such
finer pursuits. So when a person goes even higher than the plane of
spirituality and of Divine inspiration, he enjoys a state of bliss. When
the sun shines the moon becomes dim. Thus the renunciation
necessary for the attainment of Love of God is not obtained by
destroying anything; it comes naturally, just as in the presence of an
increasingly stronger light, less intense lights become dimmer and
dimmer until they fade away completely.

So when the Love of God springs up, the love of sensual pleasures
and of the intellect fades away. The Love of God then grows into
Supreme Devotion (*Para Bhakti*). Then all the little limitations and
bondage – forms, images, rituals, books, temples, sects and
nationalities – fall away naturally from the devotee of Supreme Love.
Nothing remains to bind him. Divine Grace loosens the binding ties
of the soul and it becomes free. So in the renunciation auxiliary to
Divine Devotion there is no repression or suppression. The devotee
has no need to suppress any single one of his emotions, he only strives
to intensify them and direct them to God.

Bhakti Yoga does not say, "Give up;" it only says, "Love the
Highest." And everything low naturally falls away from him who loves
the Highest. This mighty attraction to God makes all other attractions

vanish for him. Bhakti fills his heart with the Divine Bliss; there is no place there for little loves. So the devotee's renunciation is that non-attachment (*Vairagya*) for all things that are not God, that results from the great attachment to God (*Anuraga*).

Narada says, "Renunciation means dedication of all activities, secular and sacred, to God." (*N.B.S.*, 8) The activities cannot be stopped; they can only be dedicated to God by a complete surrender of the individual soul to the Divine. This implies complete consecration of body, mind and their powers to God. The devotee's work becomes worship of God. He may live in the world, but is not of it. As Sri Ramakrishna says, "Let the boat stay on the water, but let not the water stay in the boat."

Sri Krishna teaches the secret of worshipping God through one's actions. He says,

"Whatever you do, whatever you eat, whatever you offer in sacrifice,
Whatever you give, whatever you practice as austerity,
O Arjuna, do it as an offering unto Me." (9. 27)

Sri Sankaracharya, having realized the secret of work, said, "Whatever I do, O Lord, all that is Thy worship." (*Sivamanasapuja*). All the activities of a true devotee of God are thus sublimated into worship. He is absolutely selfless and unattached to the fruits of his actions. What he renounces is not the actions but the ego, the sense of "doership" and that of "I, me, mine." Even the distinction of secular and sacred activity disappears for an illuminated soul, because every work is sacred to the devotee, as it is an expression of his love for God.

Unification

In such renunciation by dedication, there is complete unification of the devotee's soul with God, and indifference towards everything opposed to it. (*N.B.S.*, 9). The word "unification" implies this: In the devotee of Supreme Love, instincts common to all human beings merge themselves in the Divine Love and are completely unified as one. They are sublimated and their energies are redirected towards service to God through service to all beings.

There is unification in another sense also. The realized devotee has no interest of his own. He feels the woes of the world as his own, and is moved by sympathy for the sufferings of living beings, and completely forgets himself in active services. There is also unification in a still higher sense. The ego of such a person becomes identified with God and his will with God's Will.

Abandonment of other Support

The unification or single-minded devotion means the abandonment of all other support. (*N.B.S.*, 10). We seek security. We may have everything – wealth, name, fame and objects of pleasure, but still feel insecure. None of these things can give us security and eternal happiness. But ultimately we find the only security in God. All else will fail us, but the Lord never fails us. We must take refuge in Him alone. When we attain Supreme Devotion to God, then alone we realize that "He is our supreme goal, He is our support, He is our Beloved Lord, He is the witness within us, He is our supreme abode, He is our true refuge, and He is our real friend."

All the things of the world such as wealth, position, strength, scholarship, etc. lead to fear, but true renunciation alone lead to fearlessness. Saint Bhartrhari says in his *Vairagya Satakam*:

In enjoyment there is the fear of illness, in social position there is the fear of calumny, in wealth there is the fear of losing it, in fame there is the fear of humiliation, in strength there is the fear of enemies, in beauty there is the fear of old age, in scholarship there is the fear of disputants, in virtue there is the fear of traducers, in life there is the fear of death. Indeed, everything in this world is accompanied by fear. Renunciation alone is fearless. (Verse 31)

Indifference to Factors hostile to Devotion

What does this indifference, referred to in Sutra 9 of the Narada Bhakti Sutras, mean? The Love of God never ends in idleness. A true devotee will avoid only such activities as are hostile to Divine; all activities and emotions prompted by selfish desires are undesirable ones. These include unfair and harmful efforts to acquire wealth, position, name and fame, and vicious emotions such as egoism, lust, greed, anger, jealousy, cheating, pride, vanity, violence and passion. All these are hostile to Divinity, because they obstruct the revelation of the Supreme Devotion, that is God Himself. The true devotee is indifferent towards these undesirable actions and demonic qualities. He turns away naturally from all undesirable actions and reactions. With the energy thus saved by the abandonment of undesirable activities, he would vigorously perform righteous activities conducive to devotion. These activities may range from pious duties like worship of God to dedicated selfless service to sick and the poorest of the poor. These activities are in conformity with the Divine Will. (*N.B.S.*, 11). The devotee becomes an instrument in the hands of God. He is guided by the Divine Will. There is nothing personal in him. He performs effortlessly actions for the good of humanity. Such activities flow through him spontaneously. (*See* also the sub-title "Non-attachment" in the next Chapter).

21

Cultivation of Bhakti: Negative Practices-II

Continuing from the previous Chapter, the Negative Practices of 'Transcending Gunas'; 'Non-attachment'; 'Avoiding Vain Discussions'; 'Shunning Evil Company'; and 'Transcending Devotion with Motives'; are discussed in this Chapter.

TRANSCENDING GUNAS

Nature of Gunas

Gunas or modes of Nature are of three types: *Sattva*, *Rajas* and *Tamas*. They come forth from primordial Nature (*Prakrti*). *Sattva* expresses itself in purity, clarity and calmness. It is stainless and luminous. It produces knowledge and happiness. It enlightens the intellect. When *Sattva* is predominant, one remains serene and peaceful; he is humble, pious, generous, merciful, loving, caring and sharing; and he practices ethical discipline, and performs selfless service.

Rajas is restlessness, passion and activity. It pleases the mind and keeps alive the passions. It creates attachment or aversion toward the objects of the world. A *Rajasic* person is full of cravings and desires. He is forced to act for their fulfillment. He runs after sensual pleasures. He is egoistic. He is ever greedy and restless. Nothing gives him satisfaction. He loses his power of discrimination. He runs after name, fame and comforts. He has no peace of mind. He is attached to actions and their fruits. He is motivated by selfishness and binds himself by the effects of his selfish actions.

If intense *Rajas* takes a Sattvic turn it can thus be transformed into a creative and constructive force.

Tamas is ignorance and inertia. It creates ignorance and destroys the sense of discrimination. It creates delusion and thus leads to inaction. A *Tamas*ic person is thoughtless and ignorant. He is negligent and indolent. He forgets everything. He identifies himself with the body, and fights with people if they speak ill of him. He has no sense of balance. He remains in a state of sleepiness and lethargy. He is controlled by negative emotions. He is depressed, dependent and helpless.

Gunas are in a state of constant interaction, and a person's mood changes whenever one guna prevails over the other two. It is the dominance of a particular guna that determines the nature of a person.

Lord Krishna, in the *Bhagavad Gita*, points out that all the three Gunas are bonds that imprison the soul within the body and prevents it from knowing the Atman, its true nature. *Tamas* is the bondage of sloth, stupidity and cowardice. *Rajas* is the bondage of lust, greed and selfish activity. A devotee should overcome *Tamas* and *Rajas* by means of *Sattva*. Without establishing *Sattva* as the predominant quality, progress in spiritual practice cannot be achieved. In the final stage by cultivating one-pointed devotion he should transcend even *Sattva*, as *Sattva* also binds one by making him to search for happiness and worldly knowledge.

Therefore, the devotee must overcome the Gunas and become free from bondage by the practice of discrimination. He must remind himself that it is the Gunas alone are the agents of action and not he. He must not identify himself with the actions, and not get attached to their fruits.

God is beyond the three Gunas. None of the three Gunas can reach God. They are like robbers who cannot come to a public place for fear of being arrested. Here is a story :

Once a man was going through a forest. Then three robbers fell upon him and robbed him of all his possessions. One of the robbers said, "What is the use of keeping this man alive?" So saying, he was about to kill him with his sword, when the second robber interrupted him saying, "Oh, no! What is the use of killing his? Tie him hand and feet and leave him here." The robbers tied his hands and feet and went away. After a while the third robber returned and said to the man, "Ah, I am sorry. Are you hurt? I will release you from your bonds." After setting the man free, the thief said, "Come with me. I will take you to the public highway. After a long time they reached the road. Then the robber said, "Follow this road. Over there is your house." At this point the man said, "Sir, you have been very good to me. Please come with me to my house." "Oh, no!" the robber replied. I cannot go there. The police will know it."

The world itself is the forest. The three robbers prowling here are *Sattva*, *Rajas* and *Tamas*. It is they that rob a person of the Knowledge of Truth. *Tamas* wants to destroy him. *Rajas* binds him to the world. But *Sattva* rescues him from the clutches of *Rajas* and *Tamas*. Under the protection of *Sattva* the person is rescued from anger, passion, sloth, lust and greed. Further *Sattva* loosens the bonds of the world. But *Sattva* also is a robber. It cannot give him the ultimate Knowledge of Truth, though it shows him the road leading to the Supreme Abode of God. Even *Sattva* is away from the Knowledge of Brahman. *Sattva* is the last step of the stairs. Next is the roof. As soon as *Sattva* is acquired there is no further delay in attaining God. One-step forward, God is realized.

NON-ATTACHMENT

One's attachment to the objects and pleasures of senses arises out of one's own egocentric desires. There is no limit to such desires. The pleasures derived from sensual objects are momentary. They are followed by anxieties, worries, disappointments and painful experiences and sufferings.

There are also attachments to living beings such as parents, spouse, children, relatives and friends. They need not be renounced, but our relationships are to be purified by changing the attitude of our love toward them. Our love for an individual becomes an attachment when it is motivated by our expectation of selfish gratification through that individual. Transform that conditional love into unconditional love. Nothing–not even one's own body–belongs to us. Everything belongs to the Lord. Give up the imaginary sense of "I and Mine."

Attachment to objects exists in the mind. So the renunciation of lust and greed must not be merely physical but be mental as well. In the *Bhagavad Gita* we read:

> "He who, restraining the organs of action, sits thinking of the objects in mind,
>
> Is of deluded understanding and is a hypocrite." (3. 6)

Some psychologists call this 'repression', that, according to them, creates complexes. So they advocate expression, that is, enjoyment of sense pleasures. But it is not the remedy. Thirst for sensual enjoyment knows no satiation. The mind continues to desire, but the senses are no longer able to satisfy, as their capacities for enjoyment are limited. Hence this leads to frustration and complexes. What, then, is the real remedy? Control the senses by the power of your will and direct your thoughts to God. "Detach and attach" is the mystic formula for spiritual progress. A devotee must repeatedly detach himself from the objects of the world and become increasingly attached to God.

Most of us are not only attached to what we have, but also to what we do not have. The biggest hurdle is to be free from attachment to things that we do not have. When we are not attached to what we do not have, then whatever we have a greater value, and we live in freedom.

Once there was a great king who earned his own living. Although he had many valuable things, after he had finished taking care of his subjects, at night he would copy the scripture. The next day he would have his servant go and sell what he had written. He lived on whatever he earned in this way.

One day the servant received a letter from his family begging him to come back home. There was some emergency situation. So the servant told the king that he must go home and that he needed his salary; he had not been paid for a long time.

The king told him, "I don't have any money right now. You have to wait for some time."

So the servant waited. But he kept reminding the king, because he was still getting letters from his family, asking him to return. The king kept on refusing, saying that he had no money. Finally three months later, the king gave him two rupees (less than fifty cents). The servant asked, "Two rupees! Is that all?"

The king replied, "This is from my earnings. These two rupees will make you the richest man, because I'm not attached to what I have in this palace. I'm not also attached to anything that I don't have. This money is very pure. So take it."

What could the servant say to his own master? He left with the two rupees. He had been working for the king for so many years. He wondered how he could go home without any gift for his family. By the roadside someone was selling pomegranates, so he bought some, and got many more than he expected for his two rupees.

As he traveled on, he passed through a country where the queen was very ill. The physician had said that she must have the juice of pomegranates, but in that country they did not grow them. So the king issued a decree: "If anyone can bring even one pomegranate seed, I will give him one thousand rupees for it."

When the servant heard the announcement, he immediately went to the palace with all his pomegranates. The king took all and as promised gave him one thousand rupees for each seed.

Obviously, the servant was now very rich, and when he returned to his family they were full of praise for his great qualities. He had been serving for many years, and they felt sure that the king had given him all this wealth because he was so pleased with him.

All this came from just two rupees earned without attachment. This is why it is said: "become free." Through freedom you become indifferent to pain or pleasure. Pain and pleasure are relative to one another. Kabir said:

Become free from both virtues and sins, because if you have virtues,
When you finish them, then there are sins.
And if there are sins, there will be virtues.
But sins and virtues become the cause of birth and death.
As you become free from both virtues and sins,
You become free from both pain and pleasure.
In turn you become free from both birth and death.

When you become free from everything, you gain patience to allow
the Lord to do every thing through you. Patience is not something
you force upon yourself. It is natural, a miracle that takes place within
through God's Grace. Patience brings about serenity and tranquility.

Inner detachment or dispassion (*Vairagya*) springs from sincere,
wholehearted aspiration to realize God. When the heart aspires, the
senses do not function in the manner they did before. They cooperate
with the heart's aspiration. Dispassion is mainly a state of mind, non-
attachment. Through reasoning you get dispassion. When you know
that tuning your life to Divine Love is the only way for real happiness
and peace. Then naturally the external objects do not attract you.
And it is easy for the mind to be turned toward Love for God. Without
detachment higher levels of devotion cannot be experienced. There
are two levels of detachment or dispassion: Lower Dispassion and
Supreme Dispassion.

Lower Dispassion has four stages:

1. Striving Stage. In this stage a devotee develops dispassion by
   thinking of the evils and fleeting nature of sense enjoyments.
   The wealth and glory of the world are repugnant, once the mind
   has tasted the joy of Divine Devotion.
2. Differentiating Stage. A devotee begins to discover the
   changes that have occurred in his inner life – the gradual waning
   of the subtle desires for objects, and the emergence of greater
   control over senses.
3. One-sense Stage. As dispassion grows and the devotee
   masters the senses, then the mind is the only "sense" that remains
   to be mastered.

4. Controlled Stage. At this stage, the mind is also controlled, and the devotee enjoys the increasing bliss of devotion.

As devotion grows, the devotee, by Divine Grace, finally renounces the mind-stuff itself and attains Supreme Dispassion. In Supreme Devotion, the devotee has no awareness of anything except God. He is ever established in God-consciousness.

When the young sage Sukadeva was going from one place to another, followed by Vyasa, his father, they had to pass by the side of a river where women were bathing without any cloth on their bodies. They were disporting themselves in water. Sukadeva passed by without turning this way or that. His gaze was turned inward and he was unmindful of his body and the surroundings, and the women did not take notice of him. Whereas when Vyasa came near the river, the women at once covered themselves with their clothes. This shows Vyasa had not Sukadeva's dispassion. The one was dead to the world, and the other was not. Sukadeva had the eyes of a child, and his father Vyasa had the eyes of an adult. Sukadeva had the sight of oneness; Vyasa had the sight of duality.

AVOIDING VAIN DISCUSSIONS

In various branches of secular learning, discussions and debates are necessary to settle differences, to ascertain the validity of a theory, and to widen the horizons of intellect. But in the realm of spirituality, the intellect is a mere means to an end. If properly cultivated under the guidance of a Guru it blossoms into intuition.

Discussions with the Guru to understand the spirit of the scriptures are fruitful, but vain discussions and arguments that lead a devotee no where should be avoided. This is why *Katha Upanishad* says, "This intuitive intellect cannot be gained by argumentation," and the *Brahma Sutras* also state, "The Truth (of Brahman) is not sustained by argumentation."

Vain discussions and debates tend to create a tense situation. One becomes more interested in defeating the opponent than in establishing a fact. A heated discussion leads to wrangling and even fighting. There is a saying, "With every discussion, the fire of animosity is kindled more and more."

A devotee should not waste his time in vain discussions. (*N.B.S.*, 74). He should conserve his mental energy for the practice of *Japa* and other practices that promote purity of mind.

Faith that unfolds in a devotee does not contradict the growth of a healthy reason, but allows reason to be transformed into intuition. It is intuition, not mere intellectual speculation, that leads to Enlightenment.

There are also other reasons for avoiding vain discussions. (*N.B.S.*, 75). These include:

1. Everything can be studied from different points of view. For example the concept of celibacy can be studied from physiological, psychological, moral, ethical and spiritual points of view. So discussion multiplies itself.

2. Aspirants are in different stages of evolution. What is meant for an advanced aspirant may not be applicable to a less advanced one. While the advanced devotee transcends all laws of society, an aspirant must adhere to them in order to rise beyond them.

3. Transcendental matters such as the nature of God, Maya, Brahman, Immortality and Liberation cannot be described adequately by one's rational faculty. Therefore, there is room for diverse viewpoints in dealing with these matters.

4. The conclusion reached by one's reason is not final, because reason is a limited instrument, and as reason is purified one's viewpoint changes accordingly.

5. As the horizon of knowledge advances, viewpoints change. Many scientific conclusions and theories that were valid some years ago have no validity today. Mystic truth like the existence of God cannot be established by argumentation. It is based upon one's inner subjective experience, i.e., Self-realization through Supreme Devotion or Meditation.

Since there is plenty of room for diverse viewpoints on spiritual questions, a devotee should not be intolerant towards others' views. Yet he should adhere to the view that best serves his spiritual evolution. This is why Sri Sankaracharya, when he heard an aged person struggling with Sanskrit grammar, said, "O Dull-witted One! Sing the Name of Govinda (Lord). When death knocks at your door, your grammar recital will be of no help. Therefore, take refuge in the Lord and sing the Name of Govinda."

SHUNNING EVIL COMPANY

Evil company means association with evil-minded people, people who have a negative influence on you. It also refers to association with all objects of temptation. An aspirant should avoid evil company, especially at the early stages (*N.B.S.*, 43). He should not justify evil association by saying that God is in everyone. Even though God is in the scorpion, one must still avoid its sting. Similarly, although God is everywhere, one must avoid evil association at all costs.

An aspirant should not also test his advancement by exposing himself to evil association. "The young plant," says Sri Ramakrishna, "needs to be hedged around to protect it from being eaten up by stray animals."

Why should the evil company be shunned? Evil company must be avoided, because it nourishes the mental impurities of lust, greed, anger, delusion, forgetfulness, and loss of reason, and leads to complete ruin. (*N.B.S.*, 44). "If, through want of vigilance, the mind deviates an iota from its aim by extraversion, then like a ball dropped on the first rung of the stairs falls down and down to the bottom," says Sankaracharya in his *Vivekacudamani* (Verse 325), a treatise on Vedanta philosophy.

Evil company brings into one's mind a steady flow of sensual thoughts and they develop in time irresistible lust for the particular sense object, demanding a quick and easy gratification. When this is not readily fulfilled, the individual develops anger toward the obstacle that stands between him and his gratification. The intensity of the anger that arises in his mind is directly proportional to the strength of the lust ungratified. When anger mounts up, delusion sets in. At this stage in the madness of his passionate self-deluding anger, he loses all his memories of his past experiences. When memories are lost and the discriminative intellect is robbed, he becomes a brute and thus reaches the state of his total ruin all by himself.

The above effect of evil association is based upon the following verses from the *Bhagavad Gita*:

> "By constantly dwelling upon objects, one develops
> Attachment to them. From attachment arises desire;
> From desire anger is born. From anger arises delusion.
> From delusion loss of memory; from loss of memory one loses reason,
> And from loss of reason one heads toward destruction." (2. 62-63)

In the beginning the impressions of the mind, propensities for lust, greed, anger, hate and the like may be easily checked; but once agitated into a storm through evil company, they become a veritable ocean difficult to be crossed. (*N.B.S.*, 45). So an aspirant should, from the beginning, always keep the senses under the control of the intellect and remain calm and contented. He must avoid becoming a prey to the pernicious influences of the evil-minded people and their association.

In order to understand the thought-waves of lust, anger, greed and the like, one must have an insight into the cycle of the mind. The subtle impressions (*Samskaras*) in the unconscious mind become subtle desires (*Vasanas*) through external association. The subtle desires manifest in the subconscious mind in the form of slight inclinations. If they are further stimulated by imaginings, they become thought waves, which in turn become strong desires or cravings. Thus the subtle impressions develop into the thought waves, and result in actions and experiences.

The subtle impressions of the unconscious can be either negative or positive. Negative impressions are based on ignorance, egoism, attachment, aversion and fear of death. They are painful ones and, when manifested, promote increasing bondage to worldliness. Positive impressions, on the other hand, are pleasant ones, and arise from spiritual experiences, elevated thoughts and Divine virtues.

An aspirant on the Path of Devotion promotes the development of pleasant impressions by practicing the disciplines of renunciation, dispassion, worship of the Lord, listening and singing the Divine Name and seeking good associations. This practice must continue until the seeds of subtle impression are scorched by the fire of Divine Love.

Sometimes one may find the association with objects of temptation unavoidable. Then one can visualize God even in the midst of evil temptation. For example an aspirant may be compelled by circumstances to mix with persons of opposite sex in the course of performing some unavoidable duty. He may then ward off undesirable thoughts by trying to see in the other sex embodiments of the Divine. The Tantric scripture advises such people to see Lord Siva or Rama in all males and Goddess Parvati or Sita in all females.

The cases of realized Sages are different from the aspirants. Such Sages have gone beyond all Gunas, all pairs of opposites like good and bad, and holy and evil and such others. As the *Upanishads* say, they see the whole world as God alone. They make no distinction between the sinner and the saint. They see God in all.

TRANSCENDING DEVOTION WITH MOTIVES

There are three kinds of devotion to God with motives. In the *Bhagavad Gita*, they are described as follows:

(1) Devotion to God when one is in distress
(2) Devotion to God for fulfilling earthly desires
(3) Devotion to God for knowledge

All these three kinds of devotion are devotion with motives. That is, one is devoted to God for the sake of some selfish purpose. Such devotion is a lower form of devotion. Of course it initiates the devotee to establish a channel of communication with God and seek his

Grace in fulfilling his desire. But he should not stop at this level. He should intensify his devotion to reach the highest stage of devotion.

Supreme Love, however, is attained only after the devotee has transcended devotion with motives. When Supreme Devotion is attained, the devotee loves the Lord for love's sake without any selfish motive, just as a devoted mother loves her child with all her heart and soul without expecting any return. The Supreme Devotee's love is a pure love, and finds an outlet in service. The highest love manifests in serving the Lord in the attitude of either eternal servant or mother. An eternal servant serves with reverence and dedication and the mother with devotion and dedication.

22

Cultivation of Bhakti: Positive Practices-I

In continuation of the previous chapter, the Positive Practices of
'Preparatory Disciplines,' 'Cultivation of Virtues,' 'Seeking Holy
Company' and 'Listening and Singing' are discussed in this Chapter.

PREPARATORY DISCIPLINES

Certain Preparatory Disciplines are necessary for developing the
Divine Love. Discipline does not mean somebody ordering you
around, "Do this, do that...." It is just a question of focus, of watching
what arises and what subsides, gaining will power to establish control
over craving and other negative emotions, and to cultivate Divine
virtues. According to the sage Ramanuja, the expounder of the
Vaisihtadvaita (Qualified non-dualism) School of Philosophy, the
following are the preparations for getting intense Divine Love:

1. Practice (*Abhyasa*): continuously remembering God
2. Discrimination (*Viveka*): discrimination of food
3. Longing (*Vimoka*): intense longing for God
4. Truthfulness (*Satyam*)
5. Straightforwardness (*Arjavam*) or honesty
6. Work (*Kriya*): doing good to all
7. Well-wishing (*Kalyana*): wishing the well-being of all
8. Compassion (*Daya*)
9. Non-injury (*Ahimsa*)

10. Charity (*Dana*)
11. Cheerfulness (*Anavasada*)

The above disciplines are described below.

Practice

Sustained effort has to be made to achieve steadiness of the mind. It should always think of God. Indeed it is a hard task, but it can be achieved by persistent practice. What we are now is the result of past practice. Thinking of the sense-objects we have become the slaves of the world. We should go the other way, think of God. Let the mind think of God alone. When it tries to think of anything else, turn it back to God. As oil poured from one vessel to another falls in an unbroken line, so should the mind think of God continuously. This practice should be made not only for the mind but also for the senses. "Instead of hearing foolish things," says Swami Vivekananda, "we should hear about God; instead of talking foolish words we should talk of God; instead of reading foolish books we should read good books which tell of God." (*Religion of Love*, pp. 9-10)

Music is a great aid to this practice of keeping God in memory. The Lord said to Narada, the great teacher of Bhakti, "I do not live in heaven, nor do I live in the heart of the Yogi, but where My devotees sing My praise, there am I." Music has such a tremendous power over the human mind, that it brings it to concentration in a moment. Even dull, brute-like human beings, and even animals become charmed with music.

Discrimination of Food

The second discipline is discrimination of food. Food contains all the energies that go to make up the forces of our body and mind. Some kinds of food like meat and alcoholic drinks have a tremendous adverse effect on the mind. They are exciting and make the mind running all the time. Foods like meat and fish are also impure.

According to Ramanuja, food (*ahara*) becomes impure for three reasons.

1. *The Nature or Species of the Food*

All exciting foods should be avoided; meat, for instance, is by its very nature impure. You can only get it by taking the life of another. You get pleasure for a moment, but another creature has to give up its life to give you that pleasure. A devotee should avoid not only meat but also all exciting foods, such as onion, garlic and stale food.

2. *The Person from Whom it comes*

The person from whom the food comes is important. This is very intricate to Western minds. This is rather a mysterious theory of the Hindus. Each person has a certain aura around him, and whatever he touches, a part of his character, influence, is left on that thing. So who touches our food when it is cooked and served is important. A wicked or immoral person must not touch it. A devotee must not dine with wicked people, because their infection will come through them.

3. *The food to be free from dirt, etc.*

The food must be free from dirt, dust and all such things. Food articles brought from the market should not be cooked unwashed. The mucous membrane is the most delicate part of the body, and the salivary secretion conveys all tendencies very easily. Its contact is not only offensive but also dangerous. So we must not eat the food that has been partly eaten by someone else.

When these three things are avoided in food, the food becomes pure; pure food brings a pure mind, and a pure mind a constant memory of God. The *Chandogya Upanishad* says,

> "Through purity of food comes purity of mind,
> Through purity of mind comes a steady memory of Truth,
> And when one gets this memory,
> One becomes free from all knots of the heart." (7. 26. 2)

In this passage the word 'food' means everything that comes in contact with the senses. We must have pure food for the eyes, the ear and the organs of touch, smell and taste.

Sri Sankaracharya, exponent of Monistic Vedanta, takes a quite another view. The Sanskrit word '*ahara*' for food is derived from a root that means 'to gather;' hence it means that which is gathered in. He says, "when food is pure the mind will become pure." This means that we must avoid certain things in order to keep the mind pure.

1. *Attachment.*
The first thing to be avoided is attachment. We must not be extremely attached to anything excepting God. See every thing, do every thing, touch every thing, but be not attached. As soon as the extreme attachment comes, a person loses himself; he is no more a master of himself; he is a slave.

2. *Jealousy*
The second thing to be avoided is jealousy. There should be no jealousy in regard to objects of senses. Jealousy is the root of all evil, and most difficult to conquer. The sage Tiruvalluvar in his great ethical text *Tirukural* says,

"The envious need no other foes; their envy is enough." (165)
"None has gained through envy, nor the unenvious ever lost." (170)

3. *Delusion*
The third thing to be avoided is delusion. We are always taking one thing for another and acting upon that, and the result is that we bring misery upon ourselves. We take the bad for the good, and plunge into it immediately and find that it gives us a tremendous blow, but it is too late. Every day we commit this mistake, and we often do it all our lives. When the senses work in the world without being extremely attached, without jealousy or without delusion, such work is called 'pure food,' according to Sankaracharya. When the mental food is pure, the mind is able to take in objects and think about them without attachment, jealousy and delusion. Then the mind becomes pure, and when the mind is pure, the memory of God stays in that mind.

Both the above two explanations are true. It is only when we take care of the real material food that the mental food will become pure. It is very true that the mind is the master, but most of us are bound by the senses. We are controlled by matter, and so long as we are controlled by matter, the purity of the material food is essential. So we have to follow Ramanuja in taking care about food and drink. At the same time we must take care about our mental food as well. Then our spiritual self will gradually become stronger and stronger, and the physical self will be less and less assertive. Jumping at the highest ideal is dangerous. We are bound down to the world; we have to break our chains slowly. This is called the 'discrimination of food.'

Freedom

The third discipline is freedom from desires. To love God one must get rid of extreme desires; desire nothing except God. The objects of senses and the world are good so far as they help us to attain the Highest. The world is only a means to an end, and not the end itself. If this were the end, we should never die. But we see people every moment dying around us. Yet foolishly we think we shall never die, from that conviction we come to think that this life is the goal. This notion should be given up at once. The world is only a means to perfect ourselves. So the spouse, children, money and leaning are good so long as they help you toward God, but if they have ceased to be that, they are nothing but evil. If the family helps you toward God, it is a good family. If money helps you to do good to others, it is of some value, if not, it is simply a mass of evil.

Truthfulness

Truthfulness is a basic rule of spiritual life. To become a devotee of God we have to live a life of truth. We must be guided by our inner voice called conscience. Thought, word and deed must be perfectly true. He who is true, unto him the God of Truth comes. Kabir says,

"There is no austerity greater than Truth.
There is no sin greater than falsehood.
The Lord abides in the heart of one who is devoted to Truth."

We also read in the ancient ethical Tamil text *Tirukural,*

"Lie not against your conscience
Lest it burns you." (293)
"Truthfulness in thought and word
Outweighs penance and charity." (295)
"All lights are not lights; to the wise
The only light is truth." (299)

Honesty

A devotee should be honest. There is no place for deceit, crookedness or falsehood in the quest for God. God wants neither your wealth nor your intellect; He wants a pure, truthful heart shorn of deceit and hypocrisy. There is nothing to hide from Him. He is the all-pervading and all-knowing Spirit. Our hearts must be as pure as the white snow and as clear as crystal. We must cultivate our hearts to become as guileless as that of children. The ancient Saint Tiruvalluvar says in *Tirukural,*

"Do not do what the wise condemn
Even to save your starving mother." (656)
"Better the poverty of the wise
Than wealth got with infamy." (657)

Work (Kriya) or Duty

The next discipline to be practiced is doing good to others. The thought of God will not come to the selfish person. The more we do good to others, the more our hearts will be purified. God dwells in pure heart. "The heart is not impure by itself;" says Swami Cidvilasananda, "It is sublime. It holds the flame of God. But the impressions of past actions and ravages of the six enemies – anger, lust, pride, jealousy, delusion and greed – encase the heart like a hard shell covering the kernel of a coconut." How do we get rid of this shell and enter the pure heart? The only way this can be done is through spiritual discipline.

According to the Hindu scriptures there are five kinds of duties called five worships.

1. *Worship:* This is one form of devotion to God. (For details, *See* the next Chapter). In addition to worshipping God, the Guru is also to be worshipped, as the Guru is a manifestation of God.

2. *Study:* A devotee must study every day the scriptures. This is a duty to the ancient sages. The study of scriptures is an important part of spiritual discipline. He must train his intellect through regular studies and deep thinking on what is studied. He must reflect on what he studies and make it a part of his own.

3. *Duty to Ancestors:* This involves offering oblation to ancestors.

4. *Helping Fellow beings:* One has no right to live for oneself. A householder is looked upon as the mainstay of society. *Manu Smrti* declares:

> "Just as all living beings depend upon air for their existence, so also people belonging to other stages of life depend on the householder for their sustenance." (3. 77)

A householder must provide shelter and food to the poor and homeless. He has no right to cook food for the family only. Tiruvalluvar, in *Tirukural*, says,

> "To eat alone what one has hoarded
> Is worse than begging." (229)

The family should have what remains of food after giving to others. In India it is a common practice to give the first fruit and other products to God. This practice makes the family unselfish. The Hebrews in olden days used to give the first fruits to God. The first of everything should go to the poor; we have only a right to what remains. The poor are God's representatives.

5. *Protecting Animals:* Everyone has a duty to protect animals. It is diabolical to say that all animals have been created for human beings to be killed and used in any way they like. "It is the devil's

gospel, not God's," says Swami Vivekananda. One portion of the food we eat belongs to the animals. They should be given food every day. There ought to be hospitals in every city for the animals like horses, cats, dogs and cows and they should be fed and taken care of. Note what Tiruvalluvar states forcefully in *Tirukural,*

> "He only lives who is kin to all creatures
> Deem the rest dead." (214)

The above five duties are called the great five sacrifices (*pancamaha yajna*). All these duties are to be performed not as drudgery but in a spirit of service and dedicated to God. Then they become a form of worship and do not lead to bondage. Vedanta aims at integrating duty, service and worship. All these activities take us toward God. As Sri Krishna tells Uddhava in *Bhagavatam,*

> He who worships Me (Lord) constantly and steadfastly through the performance of his duty, knowing Me as the supreme Goal, becomes endowed with knowledge and realization and soon attains to My Being. All duties, performed with devotion to Me, leads one to liberation. This is the way of blessedness. (11. 18. 44, 47)

Well-wishing

Wishing well of others is another quality of a devotee. With a loving heart we must pray to the Lord for the well-being of others, for the peace and welfare of the world. There should be no feeling of grudge or hatred against others, even if they have done us harm. Wish for them all that is good. This is the spirit of a true devotee. Buddha said, "Hatred ceases not by hatred but by love."

Compassion

Compassion is the next. God is love, and compassion flows from it. He is mercy and compassion personified. As we seek to realize God, we too will have to cultivate these qualities and become an ocean of mercy and love. Let our heart be as broad as the infinite sky.

Non-injury

This is a highest virtue. Do not cause any injury to any creature by thought, word and deed. Tiruvalluvar in his *Tirukural* exhorts, "Do not do to others what you know as hurt yourself." (316). Ill will, hatred, anger and malice cannot stand before this powerful virtue. It is a special attribute of the soul. The cultivation of this virtue generates the feeling of universal brotherhood and cosmic love.

Charity

"The only gift is giving to the poor; all else is exchange," says Tiruvalluvar in his *Tirukural*. Spontaneous charity with a pure heart for relieving the suffering of the poor is a potent means of growing spiritually. Charity destroys mean-mindedness and miserliness. The more we give, the more we get. This is the Divine Law. Give the last bit of bread you have, even if you are starving. "You will be free in a moment," says Swami Vivekananda, "if you starve yourself to death by giving to another."

Cheerfulness

Cheerfulness is an essential quality for a devotee. There is no room for depression on the spiritual path. Always have hope and never give up. Many forces may assail you and pull you down. But you should not lose hope or give way to depression. Brave the storm and proceed in spite of the difficulties and adverse conditions. With cheerful perseverance and optimistic zeal, relying on God's Grace, proceed with your practice. By being cheerful and smiling you shall be nearer to God.

God is not reached by the weak. You have infinite strength within you. At the same time you must avoid excessive merriment. Excessive merriment is always followed by sorrow. Do not run from one extreme to another. Let the mind be cheerful but calm.

CULTIVATION OF VIRTUES

Most of the self-styled 'devotees' think that by worshipping God regularly and performing other prescribed ceremonies and by making offerings to God, they have discharged their devotional duties. After the worship, they have no scruples to tell deliberate lies or cheat fellow beings. They also have no control over the senses. Yet they regard themselves as devotees. Le them hear what the Divine Sage Narada has to say:

> "The devotee should cultivate harmlessness, truthfulness, purity, compassion, faith and other such virtues." (*N.B.S.*, 78)

The followers of the Path of Devotion, like the followers of other Paths of Yoga, have to cultivate ethical virtues and lead a moral and ethical life. Ethical life is a basic requirement of spiritual pursuit. All the scriptures emphatically exhort the spiritual seekers to attain purity of mind and heart by cultivating ethical virtues. Only those who are pure in heart shall see God.

The *Katha Upanishad* declares:

> "He who is intelligent, who has control over mind, and who is ever pure Reaches the goal from which he is not born again." (1. 3. 8)

In the following three verses of the sixteenth Chapter of the *Bhagavad Gita*, Lord Krishna states the Divine Virtues possessed by godly people:

> "Fearlessness, purity of heart, steadfastness in knowledge and Yoga, Charity, control of senses, sacrifice, study of scriptures, Austerity and straightforwardness,
> "Harmlessness, truthfulness, absence of anger, renunciation, serenity, Broad-mindedness, compassion toward beings, non-covetousness, Modesty, absence of fickleness,
> "Vigor, forgiveness, fortitude, purity, absence of hatred, absence of pride – these belong to one born in a Divine state, O Arjuna." (16. 1 – 3)

Lord Krishna also described Divine nature in detail in the state of steady wisdom (*Sthitaprajna*) in the verses 55 – 72 of the Second Chapter, in the state of grand devotees in the verses 1 – 20 of the

Twelfth Chapter, and in the state of a Sage who has transcended the Gunas (*Trigunatita*) in the verses 22–26 of the Fourteenth Chapter of the *Bhagavad Gita*.

The ancient sage Patanjali in his *Yoga Sutras* has laid down the ethical virtues under *Yamas* (Restraints) and *Niyamas* (Observances). The five *Yamas* are: non-injury, truthfulness, non-stealing, continence, and non-covetousness (2. 30). The five *Niyamas* are: purity, contentment, austerity, study of scriptures and surrender to God (2. 32). He also adds that these are virtues to be observed by all irrespective of class, place or age. (2. 31).

Vyasa in his *Vedanta Sutras* lays down: "Even though rituals are to be given up, one has to take up the practice of mind control (*Sama*) and sense control (*Dama*): These have been prescribed by the *Vedas* as essential limbs; so they have to be necessarily cultivated by Practice." (3. 4. 27)

Lord Buddha has given the ethical virtues under "ten virtues." Jesus Christ has given them under "Ten Commandments."

The Chinese philosopher Mencious once said, "The virtues are not poured into us, they are natural to us. Seek them and you will find them; neglect them and you will lose them." Virtues lie hidden within us. Learn to find them and take care of them. Bhartrihari, a great Indian sage, said, "O wise man, develop a regular practice of cultivating Divine virtues, for they make wicked men good, foolish men wise, enemies friendly, and invisible things visible. Divine virtues instantly turn poison into honey."

The positive virtues neutralize the effects of desire, anger, and greed – the three gates of hell – and eventually eliminate them altogether. The scientific findings indicate that most diseases are caused by mental and emotional disturbances. It is therefore necessary to cultivate attitudes and virtues that ensure a steady and peaceful mind. Mental peace and sound health are prerequirements for spiritual practice. The Divine virtues seek to remove all mental and emotional disturbances and thus to make the mind peaceful and serene.

Let us now discuss the significance of the important virtues.

Harmlessness (*Ahimsa*)

Harmlessness means non-hurting any creature by thought, word or deed. How can a devotee who feels the presence of God in all beings harm any of them? He loves all beings in the same way as he loves God. Feelings of love and compassion purify the mind.

The test of harmlessness is absence of jealousy. The real lover of mankind is he who is jealous of none. One who does not cheat anyone, whose heart never cherishes even the thought of injury to anyone, and who rejoices at the prosperity of even his greatest enemy is a real devotee.

Truthfulness

Truth is God. It means speaking what is kind and beneficial to others. Thought, word and deed should be perfectly true. God of truth comes to him who is true. One lie leads to more lies. Any effort to cover up falsehood and deception creates guilty feeling and inner conflict in the mind. Therefore truthfulness has to be practiced for keeping the mind free from such disturbances. Only a pure mind can reflect the Divinity within. "God cannot be realized without guilelessness," says Sri Ramakrishna, "Even those engaged in worldly activities, such as office work or business should hold to the truth. Truthfulness alone is the spiritual discipline in the *Kaliyuga* (Modern Age). It is very difficult to practice other austerities in the *Kaliyuga*. Truthfulness, submission to God and looking on the wives of other men as one's own mother – these are the means to realize God."

Purity

Purity is the bedrock of Devotion to God. External purity is cleanliness of the physical body and food. Cleanliness is next to Godliness. It is easy to clean and purify the body. The body is purified by refining it with pure and non-fat vegetarian diet. Meat, alcohol and smoking make the body impure and unhealthy. Internal or mental purity is more important than external purity. The dirt of ego and *Tamasic* and *Rajasic Gunas* makes the mind impure. The mind is purified by replacing all selfish and evil thoughts and emotions by pure and positive thoughts and emotions. The qualities that purify the mind, as given by Ramanuja, are truthfulness, honesty, sincerity, compassion, harmlessness and non-covetousness. Cultivate discrimination and dispassion. Free the mind from lust, greed, anger, and attachment and aversion to the objects of senses. Wash out all the impurities from the mind and heart by prayer and *Japa*. Lord loves a pure heart. Kabir says,

> Rama, the Lord, has possessed me, Hari, the beloved Lord has enchanted me,
> All my doubts have flown like birds migrating in winter.
> When I was mad with pride, the beloved Lord did not speak to me.
> But when I became as humble as ashes, the Master opened my inner eye.
> Dyeing every pore of my being in the color of love,
> Drinking nectar from the cup of my emptied heart,
> I slept in His abode in Divine ecstasy.
> The devotee merges with the Lord like gold merging with its luster.
> My Lord loves a pure heart.

Love itself is so pure that a pure heart alone can sustain the experience of it. Swami Muktananda wrote in *Reflections of the Self,*

> Never compromise your purity
> Perfect it within and without
> Be as radiant as the sun
> Have love as fresh as the moon
> Be as clear as spotless crystal.

The heart is precious. Lord sees only the quality of your heart. You can know God, only if you have a pure and steady heart. All virtues come from a pure heart. The heart never schemes or plots to get its own way. It never looks for its own advantage. The heart melts. It overflows. Its very nature is pure and its natural inclination is toward kindness, love, generosity, valor, compassion, forgiveness, innocence, righteousness and honesty.

Compassion

Compassion is active benevolence, a positive expression of love, as 'harmlessness' is the negative expression of it. These two form the obverse and reverse of the same coin. The statement "Do unto others as you would have them do unto you" expresses the significance of compassion. It is the love one feels for all beings of the world. It is an attitude of equality. It flows out in love to save another. As a devotee sees God in everything, his kindness is universal. Tiruvalluvar in his *Tirukural* says,

"The richest of riches is kindness; mere pelf
Even the mean possess." (241)

Faith

This means faith in God, in the scriptures, in the words of the Guru and in oneself. Faith in God is the first step to God-realization. It is the gateway to the Kingdom of God. No progress is ever possible in spiritual practice without faith. "Whether you believe in God with form or God without form," Sri Ramakrishna says, "you will certainly realize God, if you have firm faith." If you have faith in God, you need not be afraid of anything. You should also have faith in the teachings of the Guru and in the holy name or *mantra* given by him. Once you have faith you have achieved everything.

There is nothing greater than faith. Faith can move mountains. It can work wonders.

Nothing is impossible for the man of faith. Once a man was about to cross the sea. Vibhishana wrote Rama's name on a leaf, tied it in a corner of the man's cloth, and said to him: "Don't be afraid. Have faith and walk on the water. But look here – the moment you lose faith you will be drowned." The man was walking easily on the water. Suddenly he had an intense desire to see what was tied in his cloth. He opened it and found only a leaf with the name of Rama written on it. "What is this?" he thought, "Just the name of Rama!" As soon as doubt entered his mind he sank under the water.

The Vaishnava Scriptures say, "God can be attained through faith alone: reasoning pushes Him far away." One must have childlike faith and the intense yearning that a child feels to see its mother.

A boy named Jatila used to walk to school through the woods, and the journey frightened him. One day he told his mother of his fear. She replied, "Why should you be afraid? Call Madhusudana (a name of Krishna)." "Mother," asked the boy, "Who is Madhusudana?" The mother said, "He is your elder brother." One day, after this, when the boy again felt afraid in the woods, he cried out, "O Bother Madhusudana!" But there was no response. He began to weep aloud: "Where are you, Brother Madhusudana? Come to me. I am afraid." The God could no longer stay away. He appeared before the boy and said: "Here I am. Why are you frightened?" And so saying He took the boy out of the woods and showed him the way to the school. When He took leave of the boy, God said: "I will come whenever you call Me. Don't be afraid." One must have this faith of a child, this yearning. God cannot be realized without childlike faith.

Faith can be strengthened by holy association, prayer, self-purification, meditation and study of scriptures. Lord Krishna in the *Bhagavad Gita* affirms,

"The person who is full of faith, who is devoted to it,
  And who has subdued all the senses, obtains the Knowledge of Self;
  And, having obtained this Knowledge,
He goes at once to the supreme peace." (4. 39)
"The ignorant, the faithless, the doubting self proceeds to destruction;
  There is neither this world nor the other nor happiness for the doubting.
  (4.40)

The faithless and doubting person does not rejoice in this world; nor he has any chance of attaining salvation. One cannot reach the goal of life without faith. But with faith one can achieve anything, even the highest *Samadhi*. The ancient Sage Patanjali in his *Yoga Sutras* declares,

The highest *Samadhi* (Liberation) is attained gradually by the practice of unflinching faith, energy (mental strength), memory (of past Yogic practices), Concentration and intuitive vision. (1. 20)

Look at the innocent faith of Kannappa, the hunter of Kalahasthi (Southern India). He had true living faith in Lord Shiva. He daily gave flesh of wild beasts as an offering to the Lord. Shiva tested the sincerity of Kannappa one day. Tears fell down the right eye in the Shiva-linga. Kannappa was sorely moved, because for him the linga was not a mere idol; it was a living God Himself. He plucked out his right eye and fixed it in the Shiva-linga. The next day there were tears in the left eye of the Shiva-linga. Kannappa plucked his left-eye and fixed it in the linga. At once Lord Shiva appeared before Kannappa and blessed him with new eyes and immortality and eternal bliss.

Such should be your unshakeable faith as a seeker of God. You will be tested by the Lord in various ways. Even under extreme trials and difficulties you should not lose your faith. Faith is your sheet anchor. With that unswerving faith you will certainly attain the Lord.

Charity

There is no higher value than charity. The hand was made for giving. For further details, *see* under *Preparatory Disciplines,* in this chapter.

Strength

Strength means both physical strength and mental strength. God cannot be attained by the weak. A weak person cannot be able to bear the impact of the mysterious forces awakened by the practice of Yoga. It is strong mind that hews its way through thousand difficulties and the obstacles that are encountered. It is the strong will that can cut off the knot of *Maya.*

Control of Senses and Mind

As the mind functions through senses, control of mind means control of senses. Mind is fed by senses and it cannot exist without senses. Therefore, the senses have to be restrained from going toward the sense-objects, and brought under the guidance of the will. This is a basic requirement for spiritual development. Then comes the practice of self-restraint and self-denial. God-realization is not possible without struggle and without such practices. The mind must always think of the Lord. It is very hard to make the mind to think of the Lord always; but with every effort the power to do so grows stronger in us.

Endless desires and thoughts make the mind restless, weak and polluted. Turn the mind toward God through prayer, *Japa*, study of scriptures and devotional texts. Once devotion to God intensifies, it drives away base desires and disturbing thoughts from the mind. In the *Bhagavad Gita* Lord Krishna says,

"With his intellect held firm, a Yogi should attain quietude
By turning the mind away from the objects of the world
In gradual stages, by directing it to the Self." (6. 25)

Non-possessiveness

The tendency to accumulate wealth and worldly objects is very strong in us. Of course, as long as we live in the physical world, we need necessities of life for the maintenance of the body. But we are not satisfied with the necessisties of Life. We crave for comforts and luxuries. We do not, however, stop with all possible comforts and luxuries. We continue to amass wealth and possessions. There is no limit to our desire for material possessions. Apart from the far-reaching imbalances that this human instinct causes in the social and economic fields, its effect on the life of the individual is terrible indeed. Consider the time and energy one spends in the accumulation of possessions and in maintaining and guarding the accumulated things. The worries and anxieties increase more than proportionately with increase in the accumulations. Then consider the constant fear of losing some of them every now and then, and the regret of leaving them behind when one ultimately bids good-bye to this world. What a colossal waste of time, energy and mental force all this involves! Is the precious human life meant for such wastage? No one who is aware of the real purpose of life can afford to waste his limited resources of time and energy in this manner. So a spiritual seeker should not waste his life in accumulating and maintaining possessions. He should feel satisfied with what comes to him in the natural course of life. And he should not feel attached to his possessions. The danger lies not in having material possessions, but in getting attached to them or in craving more.

Contentment

Contentment is necessary for the spiritual seeker for keeping his mind in a state of equilibrium. This state of mind does not depend upon one's material possessions. A monarch can be as content as a beggar can. Our desires are insatiable and there is no end to desires. One desire leads to many other desires. The mind is, therefore, in a constant state of agitation. It is constantly subject to anxieties, worries and disappointments. Besides, we are subject to all kinds of situations and encounters, and we react to them according to our nature. These reactions involve in most cases disturbances of the mind. This state of mind is not conducive to spiritual practice. Therefore, a devotee should feel satisfied with what he has, and accept whatever happens to him.

Asceticism

Asceticism (*Tapas*) implies self-discipline. The mind is like a wild horse tied to a chariot. Imagine the body is the chariot, the intellect is the charioteer, the mind is the reins, and the horses are the senses. The Self or true you is the passenger. If the horses are allowed to gallop without reins and charioteer, the journey will not be safe. Although control of senses often seems to bring pain in the beginning, it eventually ends in happiness. If asceticism is understood in this sense, we will welcome pain; we will even thank people who cause it, since they are helping us to burn out our impurities.

A simple life free from sensual indulgence, regulated fasting, observance of silence for a day every week, chanting the name of the Lord, and selfless service to fellow men and other beings – all these constitute asceticism. It helps to discipline the mind and to develop the willpower. Self-discipline is absolutely necessary for spiritual progress. It will not obstruct one's normal life. Rather it will make one's life meaningful and peaceful.

SEEKING HOLY COMPANY (SATSANG)

Holy company means the company of Saints or a group of devotees of God singing or reading about God and His glories. The company of Saints is an essential means for awakening the soul and making it progress rapidly in its march toward the Knowledge of God. "Take refuge in some soul who has already broken his bondage," exhorts Swami Vivekananda, a great Saint of the modern age, "and in time he will free you through his mercy." It is hard to obtain the grace of a great soul, because it is hard to recognize such a one; but if a devotee receives his grace, the effect is infallible. (*N.B.S.*, 39).

It is difficult to find a spiritual preceptor. *Katha Upanishad* says,

"Many do not find opportune conditions to even hear about Him.
Even heard of Him, many are not able to know Him.
It is a wonder to find one who is able to teach the nature of Brahman.
It is a wonder to find one who has attained Realization of Brahman.
And it is a wonder to find a skilled aspirant who has been taught
By a knower of the Self." (1. 2. 7)

Lord Krishna echoes the same opinion in the *Bhagavad Gita*,

"Among thousands of persons, a few perchance strive for perfection;
Even among those successful strivers,
Only one perchance knows Me in essence." (7. 3)

How can one recognize a Saint? A Saint is one who has surrendered his body, mind and Self to God, and renounced 'lust and greed.' He is one who has attained Self-realization. He constantly thinks of God and does not indulge in any talk except about spiritual knowledge. Moreover, he serves all beings knowing that God resides in everybody's heart.

Though it is not easy to get the association of a Saint, nothing is impossible for the Lord; the moment an aspirant is ready, by

God's Grace he will get a holy association. It is only through the Grace of God, an aspirant obtains the grace of a great soul. (*N.B.S.,* 40). Sri Sankaracharya in his *Vivekachoodamani* says,

"Aspiration for spiritual unfoldment and the loving protection of a wise man are only caused by the blessings of the Lord."

The influence that a great soul sheds on the disciple is subtle and incomprehensible. The effect of saintly contact is unerring even if the transformation brought about in the seeker is slow and gradual. Through the grace of the Guru, the seeker attains Supreme Devotion and union with God. We read in *Bhagavatam,*

"Many have attained the highest illumination, not by the study of the Vedas,
   nor by the practice of austerities, but merely by loving and serving the men of God.."

Speaking of the glory of holy association, Sri Shankaracharya says,

"By good association, one attains the absence of other association. By the absence of other association, one becomes free of delusion. Freed of delusion, one becomes established in the Immutable Truth. Having established in that Truth, one attains Liberation while alive. Nothing is great in this world than holy association. Sages and Saints are moving temples of God. Their lives are living commentaries on the great scriptures. Therefore Narada exhorts, "Seek their association with all your heart and with all your energy." (*N.B.S.,* 42)

A person's temperament and outlook changes according to the company he keeps. Holy company is an invaluable means for the removal of the veil of ignorance that covers the soul for becoming aware of its immortal and blissful nature. In the words of Sri Ramakrishna, "Holy company begets yearning for God. Nothing is achieved in spiritual life without yearning. By constantly living in the company of holy men, the soul becomes restless for God." It intensely longs for God-realization, and achieves it through His

Grace. It becomes one with God when its devotion matures into Supreme Devotion.

All those, who are turned to the path of spiritual realization have received the magic touch of Saints. Even the worst of men, such as Valmiki, Paul, Jagai and Madhai, by coming under the influence of Saints, attained illumination. The contact of Saints alone is responsible for the regeneration of the fallen soul. The more an aspirant comes in contact with Saints, the greater is his progress towards the goal. No doubt the contact of a Saint awakens the soul; but to establish oneself firmly in the higher consciousness which one attains in the Saint's presence, one has to practice self-discipline and cultivate Divine virtues and deep devotion to God. One's self-effort is led to spiritual success by the magic touch of Divine Grace and the Grace of the Holy Saint.

LISTENING AND SINGING

How can one engaged in multifarious duties of worldly life develop devotion? The Sage Narada says, any individual engaged in the activities of the world can develop the love for the Lord by listening and singing of the glories of the Lord. (*N.B.S.*, 37). This is a positive method of constant remembrance of God.

According to Sage Garga, a devotee must delight in the following:

1. Listening to the Divine teachings of Sages and Saints.
2. Listening to the stories of Incarnations like Rama, Krishna and others that engender devotion in the heart and remove all mental impurities.
3. Studying devotional scriptures.
4. Singing the Names of God and His glories (*Kirtan*).
5. Repeating the Divine Name with feeling and devotion (*Japa*).
6. Delighting in good association (*Satsang*) wherever the Divine Name is praised and the Divine glories are recited.

In the Indian culture, aspirants have received inspiration through Sages who wander from place to place, telling the stories of God or explaining the teachings of the sacred scriptures. After such listening, an aspirant reflects upon what he learnt, and meditates on God. He directs all his sentiments to love and adore God and subjects his will to the Divine Will. He transforms all actions into Divine worship by developing the attitude that God is the Doer in him while he is merely an instrument in the Divine Hands. Thus he moves toward God by elevating his personality through reason, emotion, will and action. All these practices gradually purify the devotee's mind and lead him to the attainment of Supreme Devotion.

A person of worldly life has to perform various daily duties – obligatory, family, professional, social, etc. While his body is occupied in the discharge of such duties, his mind can continue to meditate on God and His blessed attributes by listening and singing. God is endowed with attributes of infinite virtue, glory, dispassion, power, prosperity and Lordliness.

Sri Ramakrishna gives several illustrations to show how an aspirant can keep his mind on God, even while engaged in various duties: a village maiden carrying water on her head, well-balanced, with her mind concentrated on the water vessel, yet at the same time gossiping with other women; a chaste wife awaiting her husband's arrival with her mind concentrated on him, while at the same time cooking meals and nursing her baby.

Bhagavatam speaks of King Parishit who was destined to die within seven days. But since he delighted in the stories of the Lord, Sage Shuka recited the stories of Lord Krishna to him. By listening to those stories, he attained Liberation. It is said that Lord Hanuman so enjoys listening to the glories of Rama that wherever the *Ramayana* is read or wherever people talk of Rama's glories, Hanuman appears there in invisible form.

The Epics of *Ramayana* and *Mahabharatam, Bhagavatam* and other puranas and scriptures like *Bible* contain glorious stories of Divine Incarnations and inspiring stories of devotees. Saints and Sages from ancient times, Alwars (realized Devotees of Vishnu) and Nayanmars (realized Devotees of Shiva) composed inspiring songs and verses in praise of Divine Name. Tulsidas, Kabir, Surdas, Guru Nanak, Lord Chaitanya, Bhagavan Sri Ramakrishna, Sri Swami Sivananda and many others expounded the glory of Divine Name.

The essence of all Yogas consists in withdrawing the mind from sensual objects and fixing it only on God. Since the mind has the property of getting the color and odor of the objects with which it is in contact, a devotee has to do all acts carefully remembering the Lord and finding delight in Godly work. All impurities of the heart are wiped out and a perfectly auspicious state is soon engendered by dwelling on the Lord continually (*Bhagavatam*, 2. 3. 10, 20-24).

The efficacy of hearing Divine discourses and singing devotional songs are referred to in *Bhagavatam* (10. 90, 49, 50):

"One who seeks to develop devotion to God should listen to narration of his deeds that wear away all enslaving Karma.... By virtue of the devotion that grows every hour and minute by listening to, singing of, and constant contemplation of, the Lord's glorious stories, the mortal human being attains Lord's abode that is beyond the range of the Lord of Death's inevitable force."

The verse 12. 4. 40 of *Bhagavatam* says,

"To a person tossed and distressed in the wild fire of various sorrows, and intent on crossing the impassable sea of worldliness there is no raft other than constant listening to, and drinking of, the excellent essence of the sportful activities of the Almighty Lord."

Both the Old and the New Testaments of Jesus recommend the spiritual practice of Chanting God's name:

"O magnify the Lord with me, and let us exalt His name together." (*Psalms*).
"Let us offer the sacrifice of praise to God continually, that is, the fruits of our lips giving thanks to His name. (*Hebrews*).

When the forms of worship like singing and listening, etc. are practiced for a long period of time, the devotee discovers that the Devotion for the Lord has taken root in him. Sage Patanjali in his *Yoga Sutras* says,

"Practice becomes firmly established when it is carried out without break for a sufficiently long period of time with perfect faith and devotion."

(1.14)

In spite of diligent practices, success ultimately depends upon the Lord's grace.

All schools of thought lay great stress upon repeating the name of God and meditating upon its meaning, as an aid to keeping our mind constantly on God. The mind that constantly contemplates upon the Lord shall get merged in Him.

23

Cultivation of Bhakti: Positive Practices-II

In continuation of the Previous Chapter, the Practices of 'Worship,' 'Puja' and 'Prayer' are discussed in this Chapter.

WORSHIP

Meaning

In Sanskrit worship is called *upasana*, which means, "to sit near" (the Deity). It means approaching God step by step until the Self attains its identity with God. Worship refers to all methods that are meant to bring the devotee face to face with God.

Stages of Worship

The temperaments of devotees vary. In relation to the different temperaments, there are four distinct stages of worship.

The first stage is Ritualistic Worship, i.e., worshipping God through an image or a picture or any other symbol. We are embodied souls and most of us have body-consciousness. Therefore image worship is naturally popular among people. (*See* "Image Worship" in chapter15. The Objects of Worship.)

The second (higher) stage is the worship with prayer (for details, *see* the latter part of this chapter) and *Japa* (for details, *see* the next chapter).

The third (higher stage) is Meditation, a constant flow of thoughts toward God. (For details, *see* the next chapter). Even in this stage, the sense of duality remains.

The final stage is to meditate on the unity of Atman with Brahman and to experience Brahman constantly. It is an actual realization of Brahman. *Mahanirvana Tantrasara*, a text on Tantra philosophy, describes the above stages of worship as under:

> Worshipping an idol or image is the first step;
> Better than this is repetition of a holy name and
> Singing of Divine glories. Better still is Meditation, and
> The last and the highest stage is to realize Brahman (14. 122)

Advanced souls do not need any symbols. They meditate on God as dwelling in their own hearts. They identify themselves with Brahman and realize Brahman, the One-without-a-second.

Purpose of Worship

Worship is done in a temple or in one's own home. The object of worship is to express the sincere devotion to the Lord and to purify the heart through the surrender of ego. During the worship the devotee should focus his mind on the form of the Lord, and think of His attributes, His Infinite Nature and Bliss. He should not think of worldly things.

Ritualistic Worship

The Ritualistic Worship is done according to the rules laid down in the scriptures. Such orthodox worship may be either external or internal (mental).

External Worship: The External Worship has sixteen stages:

1. *Avahana*–invoking the Presence of the Deity in the image
2. *Padyam*–offering water to wash the feet
3. *Asana*–offering seat for the Deity
4. *Arghyam*–offering water to glorify the Deity
5. *Achamana*–offering water to be sipped by the Deity
6. *Maduparkam*–offering a drink of honey, milk and ghee
7. *Snanam*–giving a bath to the Deity
8. *Vastram*–offering clothes

9. *Abhushana*–offering ornaments
10. *Gandham*–offering sandal paste
11. *Pushpam*–offering flowers
12. *Dhupam*–offering incense
13. *Deepam*–offering lights
14. *Naivedyam*–offering food
15. *Tambulam*–offering betel leaf, nuts, etc. for freshening the mouth
16. *Namaskara*–prayer

Offerings need not necessarily be costly things. Anything – a flower, a leaf, water or fruit – may be offered. It is the feeling of love for God that is important and not the material that is offered. The offering symbolizes the devotee's dedication and surrender of the ego.

We should worship God with the things we get from Him. All that we get comes from God. We should offer them to Him mentally. We should accept them as a sacramental thing and use them with care and devotion. By handling sanctified things we sanctify ourselves. Gradually a spirit of surrender and detachment come to stay with us.

Mental Worship

The Ritualistic Worship is also done mentally. In the Mental or Internal Worship, a devotee invokes the presence of God within his heart and mentally performs the various acts of loving service. In this worship, the devotee is not limited in his resources. He can offer all the flowers and all the treasures of the world at the feet of the Lord. Mental Worship is much more effective than External Worship.

In practice, Mental Worship means contemplation on God. This is a main spiritual practice in the Dualistic and Qualified Monistic systems of Vedanta. These systems regard God and the Self as two separate entities. The devotee at first remains the servant and God the Master. In an advanced stage, the devotee remains as the soul and God as the Soul of all souls. This attitude may serve as a stepping stone to the path of Non-dualism (*Advaita*).

While External Worship depends upon external environment and things, Mental Worship does not depend upon external conditions. Wherever you are, you can tune yourself with God within. Then meditation comes automatically through constant remembrance.

Dualistic and Monistic Worship: Monism or Non-dualism (*Advaita Vedanta*) originates from the *Upanishads*. Sankaracharya was the first philosopher to establish a formal School of Advaita Vedanta. Sankaracharya holds that the individual Self (*jiva*) and Brahman (God) are identical. Brahman is the only Reality. The universe is the projection of *Maya*, an inherent Power (*Sakthi*) of Brahman. Because *Maya* veils the Truth, the individual self misconstrues both the world and itself as different from Brahman. When the veil of ignorance is removed through intense spiritual practice, the Oneness of the Self is realized.

Ramanuja holds that the *Jiva* and Brahman are related as a part and the whole. The world and the selves are parts of the one Brahman. The Self is released from the limitations through devotion to God. The released Self has a permanent intuition of God. This view is called Qualified Monism (*Visistadvaita*). Madhva holds that the Self and God are two different entities. The world and the selves are subordinate to God and dependent on Him. This view is called Dualism (*Dvaita*).

Most of the devotees adopt a dualistic approach to worship. They consider themselves as different from God and worship Him as the Supreme Being endowed with form. They want to enter into a personal relationship with God as Master, Father, Mother, Friend or Beloved. (*See* chapter16. Attitudes (*Bhavas*) of Devotees.)

The devotees who worship God endowed with form pursue the objective method. God is the external object of their worship and adoration. They want to unite their souls with God through devotion.

The second group of devotees worship God to establish personal-impersonal connection with Him as the infinite whole and the Soul of souls. Aspirants of this type want to realize the Divine Principle as their very Self or as the One without a second. Unable to adopt the discipline directly leading to this realization, they follow a *graduated* course of Dualism, Qualified Monism and Monism.

The devotee, who follows a graduated course, attains his monistic goal by stages. Making his own soul the center of his being, he experiences in it the touch of the Divine as infinite Consciousness from which his existence is inseparable – a fact not recognized before owing to ignorance. Subsequently, with the unfolding of his spiritual powers, the devotee comes to be identified with the Infinite and to think in terms of the Infinite. He regards his being and also all other beings as manifestations of the all-pervading God. First realizing the One as connecting the many, and then the many as manifestation of the One, he finally attains to the transcendental experience of the One without a second. He thus follows the graduated path leading to the Absolute.

A rare and highly qualified small group of devotees who have attained moral perfection and dispassion follow the direct path to the Absolute. This is Monistic Worship. This transcends subject-object relationships of all kinds.

The Monistic aspirant follows from the very beginning the subjective form of spiritual practice. In this practice, the object of meditation is his own true Self that is infinite and absolute by nature. He usually meditates on the Monistic maxim, "I am Brahman" or "Thou art That." By means of this he tries to expand his limited consciousness to infinite and absolute Consciousness or Super-consciousness, rising above all limitations, physical and mental. At first he takes the help of Dualism of subject and object inseparable from relative existence, though in this he stresses the subject or Self more than the object. And finally, he attains the Absolute Divinity that transcends all duality and relativity, and all conceptions of the knower, known and knowledge. In the words of Sri Ramakrishna, he is like a salt doll dissolved in the ocean.

The mental purification attained through strenuous ethical and spiritual practices enables this type of seeker to awaken his intuitive power. He is, as it were, brought face to face with Reality, and he gains identification with It. The faculty of intuition is an innate capacity for direct experience, lying dormant in the soul. It manifests the moment the obstructions that stand in its way are removed. Spiritual practices remove the obstacles in the form of misconception. With the result, the true knowledge of the Absolute manifests itself, just as the Sun hidden by the clouds shines in all its glory when the clouds are dispersed by the wind.

The task of the direct Monistic devotee is, however, not simple. It is much harder than that of others, so he must be better equipped than others, and must possess special qualifications that will help him to march directly to the goal. Discrimination between the real and the unreal, dispassion for all enjoyments both here and hereafter, control of the mind and senses, withdrawal of senses from objects, forbearance, concentration, faith in the ideal and the path, and a yearning for perfect freedom – these are the qualifications required of such resolute seekers. When these conditions are fulfilled, the seeker is able to take up the highest form of meditation on the One without a second, and attains to the Super-conscious state of the Absolute in due course. (Source: Swami Yatiswarananda, *The Divine Life*, pp. 62-66)

PUJA (PERSONAL WORSHIP)

One of the practices in Bhakti Yoga is personal worship of the Lord. This is called *Puja*. God is omnipresent as the Cosmic Consciousness, but due to our limitations we try to perceive and worship that Cosmic Presence through an image or symbol. We should always remember that God is not really limited to just that image or symbol. We are limited, so in a way, we try to limit God.

The purpose of Puja is to slowly gather the mind for meditation. By going through all the preparation and rituals, you gently coax the mind inward. You slowly calm down all the physical and sensual vibrations. It is like putting a restless child to sleep. You cannot just simply place him in bed and expect him to sleep. You have to wash him, dress him, give him a nice drink, gently place him in bed, tell him some nice stories, and then slowly, he will drift off to sleep. It is a very practical approach.

How to perform a Puja

Puja is performed daily to the image of the Personal Deity. It is also performed in a temple by a priest for the benefit of the congregation. In the Personal worship, the image or icon or a picture of the Deity is kept in a suitable place in the home facing eastward toward the rising sun.

Take a bath in the morning. Have no selfish desires in performing Puja. Adorn the center of your forehead with holy ash (*vibhuti*) or sandalwood paste or *kumkum* (red powder). This spot between the eyebrows is the seat of latent wisdom and of concentration of the mind, and so vital for worship. Have everything ready before hand.

You begin by meditating on God as the Cosmic Being who resides in you and then request Him or Her to come and be present in the image on the altar. It is like saying, "You are everywhere, but I cannot see You, I cannot understand You. So please come and reveal Yourself through this symbol."

Sprinkle water, suggesting purification of your physical self and the environment. After invoking the Divine Presence, you symbolically receive and treat the Lord as a very beloved and revered guest. You do everything to make Him comfortable. This includes giving Him a bath, decorating Him with beautiful ornaments and garlands, burning some fragrant incense, waving a light, and offering Him some food and drink that you have lovingly prepared. After He is satisfied, you glorify Him by reciting His praises and then gently add any request you might have. This is followed by repeating His *mantra* while offering flower petals. And finally, a camphor light is waved. When the camphor burns, nothing is left. So the idea here is that when you get touched by the Divine Light, you lose yourself completely, you become one with the Light. That is what you call true realization. That is why it comes at the end of the Puja.

Then you spend some quiet time in meditation. Afterward, you offer the fruits of the Puja to the Lord and ask Him to come back and stay in your heart. You began by meditating on Him within you. Then, you invoked His presence to come out and stay on the altar. When everything is over, you ask Him to come back into you.

Finally, there is closing peace prayer. You can then distribute prasad to everyone present. Prasad refers to items including food that have been offered to the Lord and blessed during the ceremony.

The Significance of the Rituals

The rituals signify the following:

- Lighting lamps before the image of the Deity symbolizes the dispelling of the darkness of ignorance, and the illumination of the mind with the Knowledge of Brahman.
- Offering flower represents the offering of our soul to God.
- Offering fruits symbolize offering the fruits of your good and bad actions to the Lord.
- Offering cooked food means thanksgiving for the bounty bestowed by the Lord.
- When the incense is burnt before the Deity, smoke spreads through the whole room. It acts as a disinfectant. Burning of incense denotes that the Lord is all pervading, and that He fills the whole universe by His Presence. The devotee prays, "O Lord! Let the desires and tendencies dormant in me vanish like the smoke of this incense, and become ashes. Let me become stainless."
- The burning of camphor and waving it clockwise before the Deity is called arati. It symbolizes your surrender to the will of God.
- Burning of camphor denotes that the individual ego melts like camphor, and the soul becomes one with the supreme Light of lights.
- The ringing of bells during arati is to keep out other noises so that you can concentrate on prayer.
- When sandalwood is grounded into a paste, it reminds you that you should, in your difficulties, be as patient as the sandalwood. Sandalwood emanates sweet odor when it is made into a paste. So also you should not murmur when difficulties arise, but instead should remain cheerful and happy and emanate sweetness like sandalwood. You should not hate nor wish evil even on your enemy.

Mental Puja (*Manasika Puja*)

You can also do the Puja internally within the mind itself. In mental Puja, you just sit there and imagine that you are setting up an altar and doing everything that you would normally be doing in the external Puja. The Mental Puja will not be easy unless you are thoroughly familiar with the outer Puja. Without knowing how to do it outside, you will not be able to imagine doing it inside. So outside Puja is important to begin with, but Mental Puja can be done more elaborately and is more concentrated, since the mind does not get distracted by all the various objects. And because it is more concentrated, it brings more benefit. Feeling is more effective than action.

Worship Regularly

Daily worship can help you to develop a sense of closeness to God and to feel His presence more in all aspects of your life. You can worship according to the formal practices of your faith, or the Puja guidelines given here, or any form or method that appeals to you. You can even make it very simple; just light a candle, sit in front of it and say, "Lord, You are the Light; please give me Light," and proceed with your meditation. You can simplify it in any way you want, but if you want to have a little joy, make it more elaborate. It should be joyful, and not a rigid thing. You must enjoy what you are doing. It is all mainly to keep you occupied in something useful and beautiful, and to make the mind more one-pointed.

On special occasions you can perform Pujas to honor a saint or spiritual teacher. Pujas can be performed to sanctify special events in your life, such as moving into a new home or beginning a new job or to offer thanks for God's blessings. They can be done as part of your private worship or you can invite your friends to share in the celebration.

By worshipping regularly, you will form the habit of turning to God through all the ups and downs in life. You will feel His guidance and support more. It will open your heart and fill it with devotion. You will receive God's Grace and ultimately attain God-realization.

Yoga Ecumenical Service

In the Integral Yoga Institutes established by Swami Satchidananda, Yoga Ecumenical Service is performed during public ceremonies like annual Yoga retreats, conferences, ecumenical gatherings, birthday celebration of the Guru and the like. In this, God is worshipped in the form of Light. In the name of Yoga, the worship is done in a universal way. The universal symbol that could be accepted by everybody is the Light. Nobody would deny the Light. That is why a Light is placed on the altar as the main symbol for worship.

Individuals are selected to represent the major faiths and offer prayers praising God as the Light. The ceremony begins with a procession. The celebrants enter carrying lit tapers in the following order: Native American, African, Sikhism, Islam, Christianity, Buddhism, Taoism, Shintoism, Judaism, Hinduism, and Other Known Faiths. This way they proceed from the youngest to the oldest faith in a counter-clockwise manner. They gather around a central Light on the altar and jointly light It. Then take their seats.

Next, the Master of Ceremonies offers an opening invocation. Then one by one, in the same order, the celebrants rise, face the central Light, and offer their prayers. The prayers may come from the scriptures or devotional writings or the celebrant can compose one himself. The celebrant for Other Known Faiths can offer a more general prayer in honor of all the other faiths.

After the last offering, all the celebrants rise together and re-light their tapers from the main Light. As they circumambulate the altar, the congregation can join them in joyful song based on a universal or ecumenical theme. Finally, peace chants are done. There can be a closing recessional, in which case the celebrants would leave in the opposite order from which they entered.

The ceremony is simple but elegant. The atmosphere can be made more festive by incorporating traditional dress, music and artifacts. It is a reverent, joyful experience that always uplifts, inspires and fills everyone with a deep sense of our essential oneness. The Ecumenical Service helps people to see the underlying unity of the various faiths, while respecting and enjoying their diversity.

One Set of Footprints

One night a man had a dream. He dreamt he was walking along the beach with the Lord. Across the sky flashed scenes of his life. As the scenes passed by, he noticed two sets of footprints in the sand – one belonging to him and the other to the Lord. When the last scene flashed before him, he looked back at all the footprints and noticed that many times along the path there was only one set of footprints on the sand. He also noticed that this happened during the saddest and lowest times in his life. This really bothered him and he questioned the Lord,

"Lord, you said that once I decided to follow You, You would walk all the way with me, but I noticed that during the most troublesome times of my life, there was only one set of footprints. I don't understand why, when I needed You most, You deserted me."

The Lord replied, "My precious child, I love you and would not leave you. During the times of your suffering and trial, when you saw only one set of footprints, it was then that I carried you."

PRAYER

Prayer is an important form of devotion to God. Through prayer you invoke God's presence and express your innermost yearnings and deepest sentiments to Him. Prayer is the communion of your soul with God. You commune your longing with God. When we are in need of something that is beyond our capacity, we resign ourselves to God, and pray for His Grace and help. Prayer purifies our heart and makes us better beings.

Prayer, in one form or other, is essential for all kinds of aspirants. The prayer of a devotee is more expressive and emotionally charged. The prayer of a jnani, the seeker of knowledge, may not be expressed in words; but his deep inner silence itself is a kind of inner prayer.

Deep Prayer

God is everywhere, not in a particular form, but as an omnipresent awareness or power. By your concentrated sincere prayer, you are tuning your mental radio to receive that power. By tuning a radio, you are not creating music. It is already there. Your tuning merely attracts the radio waves. If your tuning is not correct or it moves a little off the particular wavelength, you will not receive the music. But the moment you find the correct wavelength, the music comes. In the same way, you will receive God's Grace only when you tune your mind to God's wavelength.

Through your prayer and devotion you are in communication with God. You make a connection. Suppose you are a bulb trying to get light, and God is the main battery. Your prayer is the connecting wire. If the prayer is a loose connection, you will not get the light. Many people just utter their prayers while their eyes roll around looking at people: 'How many people are here? Did my friend come?' That sort of prayer is just a lip service. They say, "I am Thine; All is Thine; Thy will be done," while one hand is holding the wallet in the pocket. Such prayer has a bad connection. While you pray you have to forget everything else.

In deep prayer, you forget everything. You forget your body, mind and personality. You lose yourself. Deep prayer always comes from the heart, not from the head. Flowery words and lengthy prayer are not really important. Prayer must come from the heart. Even if it is a meaningless word, God understands our innermost prayer. How does a baby communicate with its mother? By a sound that no one else can understand. Only the mother understands it. In the same way God understands our sincere deep prayer, even when we express it in a meaningless word. Always develop the heart. Wherever you go, whatever you do, let your heart be there. The sacred heart is a secret heart. You need not expose it or show it to others. Just let God know. Let it be a secret communication. Just allow the heart to melt. It is the best way to get God's Light.

Open Your heart

Once a girl wanted to be away from the sun. So she said, "Go away, sun, I don't want you." She closed all the doors and windows, sat in a dark room, and started cursing the sun: "I don't want to see you anymore. Don't come into my house."

After awhile she slowly opened the door, and the sun started pushing in. She shouted, "No! Get out!" and slammed the door in the face of the sun. Then after awhile she opened the door a little, and again the sun came in. She, then, cried, "What is this? Aren't you ashamed? Don't you have any sense of pride? I am scolding you. Aren't you offended? Why don't you stay away? Why do you want to come into my house?"

The sun just smiled and said, "That's my nature. If anybody opens the door, I just walk in. I don't wait for your invitation, and I am not offended by your scolding. All I need is an open door. If the door is open, you can't even stop me from coming in."

God is like the sun. He shines on everyone equally. He loves everyone equally. And just like the sun, God is always there outside the door of your heart. If you open that door, even a little bit, He will come shining in.

Faith and Trust

Pray to the Lord with absolute faith and trust. There is no power greater than faith. A sincere prayer will certainly be answered, but that does not mean you will necessarily be given what you want. The answer might be: "It is no good for you, my child. However much you cry, I can't give that to you because it will hurt you. Without knowing how to use it, you may hurt others also. It is dangerous." If you understand that and accept it, you prove that you have trust in God.

Do not just trust in God when you get everything. If, out of one hundred requests, ninety-nine are denied, you should still have that faith. That will eventually lead you to a higher form of prayer where you do not request anything. You simply say, "Why should I ask You? You know what is good for me, and what is not good for me. Do whatever You want. Why am I asking for something? May be You made me ask. Even the asking seems to be Your prompting. Because You wanted me to ask, I am asking. Still, I leave it to You to give it or not." This world is a Divine Plan. Every situation–adversity or prosperity–in which God places you has a meaning. Everything happens for your own good.

The great Saint Ramalingam of Southern India prayed to the Lord: "You are feeding me, I am fed. You are making me sleep, I am sleeping. You are showing me, I am seeing. You are making me happy, I am happy. And not only me – the entire universe is like that. You are the One behind every movement, every experience." If you want to have a prayer, pray to God to help you, always remembering this Truth: you are His child and he is taking care of you every minute.

Prayer and Meditation

Sincere whole-hearted prayer is meditation. You are one-pointed in your attention. When you focus your mind on a particular idea connected with God, then prayer is a form of meditation. An un-meditated prayer without full concentration will not be very powerful, and it is not surprising if such a prayer brings no satisfaction. There has to be proper attunement; the mind should be one-pointed, focused and free from selfish motives.

Every thing–Prayer

In the beginning, you may pray at a specific time, and at other times you may be engaged in some work. That is, you may be separating prayer from work. Nothing wrong in it. It is a good way to start setting time for prayer and work. But gradually it expands until the prayerful attitude pervades your entire life. In the end everything is prayer. (For details *see* chapter 28. Practice of Bhakti in Action.)

Stages of Prayer

According to Ghazali, a Muslim mystic, prayer has three stages: verbal, mental and merging with the Lord's Will. In the first stage, the devotee sings the glory of God, chants His praise and pours out his anguish of heart in melodious hymns.

Secondly when the mind becomes calm, when the outgoing senses have been restrained by sustained practices, when the mind cannot be easily affected by evil influences, prayer becomes mental. Then no physical effort is needed.

In the third stage, when the mind gets concentrated in the Divine, when it loses its outward attractions and becomes serene, devoid of desires or cravings, the prayer becomes automatic, natural and habitual; it reaches its highest stage. The Lord's Will becomes the devotee's own will; for his mind merges on God who is invoked by his prayer; there is no self-consciousness for him. He abides in the Lord. He perceives nothing external, nor anything internal. He even forgets that he is praying to the Lord and he is absorbed in the Lord's Will. But he has only one experience – his oneness with the Lord.

Kinds of Prayer

The practice of prayer is of four kinds – spoken, silent, congregational and inter-religious.

Spoken Prayer: This includes oral and group prayers and reciting devotional hymns and songs. Such prayers, when spoken with inner feeling and devotion, powerfully purify the worshipper's mind and helps him to attain a state in which he experiences God's presence.

Silent Prayer: This is a rich and liberating experience from which may spring forth many other activities. It may issue in thanksgiving, self-examination, penitence or service. In fact, the value of prayer is only to be found when a person stands before God without any reservation and in complete dedication in the fullest realization of his own littleness and of the immense power and love of God. This is fully possible only in silent prayer.

Silent prayer can be practiced at any time. It works on the subtle being. It stills the mind and establishes constant remembrance of God. Practiced regularly, it constantly recalls the soul to its source. The more the heart is cleansed through prayer, the more it is, like a clean mirror, able to reflect the image of God within. Such prayer, as Ibn Ataillah says, is "a revealer of mysteries."

Congregational Prayer: Congregational prayer is practiced in all religions in some form or other. It involves a community of devout souls. It checks egotism and self-centered living. It breaks down barriers among human beings. In the words of Mahatma Gandhi, "a congregational prayer is a mighty thing. What we do not often do alone, we do together. . . . It is a common experience for people, who have no robust faith, to seek the comfort of congregational prayer."

Inter-religious Prayer: Inter-religious prayer is an essential aspect of higher cultural and spiritual life. Respect for religions other than one's own is an essential requisite of cultural life and a fundamental necessity to the harmonious working of any decent society. In order to create a sense of unity, love and respect for one another, people must gather for inter-religious prayers from time to time. The followers of different religious faiths will be able to find a deeper unity of heart by worshipping God together. Community life may include morning and evening gatherings in common prayer accompanied by reading of selected teachings of various religions.

In the common worship of the Lord, the followers of different religions may join their hands and hearts, laying stress on the universal aspect of the Divine. It serves as a great bond uniting people of different religions and inspires them to work together for the common good in a spirit of universal brotherhood. The prayers in some ashrams like Mahatma Gandhi's ashrams in India, Sivananda ashrams and Satchidananda ashrams contain recitations from the scriptures of different religions.

In common prayer, we do not think of ourselves; our minds and hearts are tuned to God, and we strive to be attentive to His voice. This adds to the richness of the fellowship. Sincerity and humility enrich all.

Selfish and Selfless Prayer: In the initial stages of devotion to God, we are apt to pray selfishly for some boon. We crave for the fulfillment of our selfish desires. There is nothing intrinsically "wrong" with this kind of prayer, as all prayer establishes a channel of communication with God, and affirms a relationship with Him. But the prayer that seeks some selfish favor is inherently limited. The seeker does not fully open himself to God's Grace. When his prayer is not granted, his faith may even be undermined.

Essentially, prayer should be selfless as far as possible. We should first pray for the good of others, for the welfare and peace of the world and for our spiritual evolution; we should pray for the eradication of our evil qualities, for wisdom and knowledge, and for saintliness.

Tukaram, a saint of India, says that one should not pray to the Lord for the removal of small insignificant difficulties. After all, pleasure and pain are purely transitory. Pain is followed by pleasure and pleasure by pain. Pain did not exist before it came and it will not exist after awhile. Similarly pleasure did not exist before it came and will not exist after awhile. So why pray to the Lord for things that are only short-lived? It is as stupid as approaching the commander-in-chief of the army to help you pull out a thorn from your foot, when it can be removed with the help of a needle or another thorn.

Therefore, why should you want to use Divine power for gaining petty material advantages that are not going to last in any way. Why should not such great power be used for spiritual growth and enlightenment? One's devotion should be utterly selfless. True devotion is free from selfish desire.

Deeper levels of communication and love can be attained only when we learn to totally surrender to God. Prayer, then, becomes true worship. Prayer as worship is a means of purification. The *Bhagavatam* says, "If a Yogi, being deluded, makes mistakes in this life, he should burn away his sins in prayer. The Yoga of Prayer is the way of atonement." The Sufi Saint Ibn Ataillah says, "Prayer cleanses the heart of the stain of iniquity."

> *Asato Ma Sat Gamaya*
> *Tamaso Ma Jyotirgamaya*
> *Mrityor Ma Amritam Gamaya.*

> Lead me from unreal to Real,
> From darkness to Light,
> From mortality to Immortality.

This *Upanishadic* prayer is the best prayer, a prayer for the Light, for the Truth, and for Immortality. The foremost prayer of an aspirant should be for the removal of his ignorance. His goal is to realize the Truth and to free himself from the meshes of those that are unreal. His goal is to realize his essential Divine nature.

Some Prayers

Here are some examples of prayers:

Gayatri: This is an ancient Hindu Prayer. It is one of the sublime prayers:

> *Om bhur bhuvah svah, tat savitur varenyam,*
> *Bhargo devasya dheemahi, dhiyo yo nah prachodayat. (Rig Veda, 3.*
> *62. 10)*

Its meaning is:

May we meditate on the effulgent adorable Divine Being
Who has created the universe, and who removes
The darkness of ignorance. May He enlighten our intellect.

Gayatri is the "Blessed Mother" of the *Vedas*. It is an effective universal prayer. The repetition of the Gayatri *mantra* purifies our heart. The repetition of 2.4 million times of the gayatri *mantra* constitutes one purascharana. Swami Vidyaranya, the reputed author of *Panchadasi,* performed Gayatri purascharana. Goddess gave him *darshan* (vision) and granted him a boon. Swami Vidyaranya asked, "O Mother! There is a great famine in the Deccan (a part of India). Let there be a shower of Gold to relieve the immense distress of the people." Accordingly, there was a shower of gold. Such is the power of the Gayatri *mantra.*

Prayer of Jesus: 'The way of a Pilgrim' (London: S.P.C.K., 1941, pp. 19-20) describes the Prayer of Jesus as follows:

> The continuous interior Prayer of Jesus is a constant uninterrupted calling upon the Divine Name of Jesus with the lips, in the spirit, in the heart, while forming a mental picture of His constant presence and imploring His Grace. . . The appeal is couched in these terms, 'Lord Jesus Christ, have mercy on me.'

Father in Heaven. The Christian Prayer to Father in Heaven is:

> Our Father, who art in heaven; hallowed be Thy name; Thy kingdom come; Thy will be done on earth as it is in heaven. Give us this day our daily

bread; and forgive us our trespasses as we forgive those who trespass against us, and lead us not into temptation, but deliver us from evil. Amen.

Mother Mary. The prayer to Mother Mary is:

Hail Mary, full of grace, the Lord is with Thee; blessed art Thou among women, and blessed is the fruit of Thy womb, Jesus. Holy Mary, Mother of God, pray for us sinners, now and at the hour of our death. Amen.

Sankaracharya

This great Sage and architect of Non-dualistic Vedanta prays to the Lord:

I adore the Lord, the Supreme Atman, the One, the primordial Seed of the Universe, the desireless and formless, who is realized through the symbol Om, from whom the Universe comes into being, by whom it is sustained, and into whom it dissolves. (*Vedasara Siva Strotam,* 5)

Prayer of St. Francis of Assisi. The most famous of his prayer is:

Lord make me an instrument of Thy Peace!
Where there is hatred, let me sow Love;
Where there is injury, let me sow Pardon;
Where there is doubt, let me sow Faith;
Where there is sadness, let me sow Joy.

Chaitanya's Prayer. One of this great saint's prayer is:

How I long for the day when an instant's separation from Thee,
O Govinda (Lord), will be as a thousand years;
When my heart burns away with its desire
And the world without Thee is a heartless void.

Sri Ramakrishna's Prayer. Sri Ramakrishna, the God-man of nineteenth century, prayed to the Divine Mother only for love:

O Mother, here is Thy ignorance and here is Thy knowledge;
Take them both and give me only pure love for Thee.
Here is Thy holiness and here is Thy unholiness;
Take them both and give me only pure love for Thee.
Here is Thy virtue and here is Thy sin;
Here is Thy good and here is Thy evil;
Take them all and give me only pure love for Thee.
Here is Thy dharma and here is Thy adharma;
Take them both and give me only pure love for Thee.

Universal Prayer

The Universal Prayer of Swami Sivananda, a great Indian Saint of twentieth century, is:

O adorable Lord of mercy and love,
Salutations and prostration unto Thee!
Thou art omnipresent, omnipotent and omniscient,
Thou art *Satchidananda* (Existence-Knowledge-Bliss Absolute)
Thou art the indweller of all beings.
Grant us an understanding heart, equal vision,
Balanced mind, faith, devotion and wisdom.
Grant us inner spiritual strength to resist temptations
And to control the mind.
Free us from egoism, lust, greed, anger and hatred.
Fill our hearts with Divine virtues.
Let us behold Thee in all these names and forms.
Let us serve Thee in all these names and forms.
Let us ever remember Thee.
Let us ever sing Thy glories.
Let Thy Name be ever on our lips.
Let us abide in Thee forever and ever.
Om Shanti, Shanti, Shanti!

Effects of Prayer

God is the indwelling Spirit in all beings. Prayer is a means for the discovery of this Soul of all souls. Prayer takes us nearer to God. It awakens our inherent spiritual powers. Prayer humbles us. When we approach God as His child or His servant with reverence and love in our hearts, we draw close to Him.

Purification: Prayer washes off the impurities of our heart with the pure waters of spiritual emotions, corrects the defects and shortcomings and prepares our mind for the reception of Self-knowledge.

Prayer is as real as the force of gravity. It can reach a realm where reason is too feeble to enter. It can work miracles. Its magnanimous efficacy is indescribable. Its potency can be hardly comprehended without actual experience. In times of danger and calamity, mass prayer works wonders. Prayer for the departed souls bestows peace on them.

Mental Peace: If any day you feel too disturbed or tired to meditate, you can just sit down for a few minutes, and pray intensely to the Lord: "Thou art purity; fill Thou me with purity. Thou art energy; fill Thou me with energy. Thou art strength; fill Thou me with strength." (*Sukla Yajur Veda Samhita*, 19. 9). This kind of prayer will calm the mind. Prayer and meditation is the only way of finding peace.

Prayer can create such a harmonious state within us that we rise above our little ego. Then we feel the touch of the Cosmic Consciousness that lies in us. Here the deepest integration takes place. The individual consciousness remains rooted in the Cosmic Consciousness.

Once Swami Vivekananda was visiting a cattle town in Western United States. Hearing him speak of philosophy, a number of university students who had become cowboys took him at his word. When he said that one who had realized the Light was able to keep one's equanimity under all conditions, they decided to put him to the test. They invited Swamiji to lecture to them and placed a wooden tub bottom up on the ground to serve as a platform. Vivekananda commenced his address and was soon lost in his subject. Suddenly there was a terrific racket of pistol shots, and bullets whizzed past his head. Undisturbed, Swamiji continued to lecture as though nothing unusual was happening.[1]

This was spiritual poise born of spiritual illumination by which the center of gravity had come to rest in the Divine Consciousness.

Awakening Intuition

The subtle faculty of intuition is lying hidden in every one of us. "The Atman," Katha *Upanishad* declares, "is hidden in the hearts of all. It is seen only by the Seers through their pure subtle intuition." (1. 3. 12). Through cultivation of virtues, through prayers and meditation, this dormant faculty is awakened; only then does spiritual life truly begin. Then the aspirant realizes that he is the Spirit, and as the Spirit he is an inseparable part of the Supreme Spirit.

Toward true Meditation

Prayers and *Japa*, when methodically practiced, create a new harmony which purifies our thinking, feeling and willing faculties, and thus lead to true meditation. As we become purer and purer in body and mind, in thought, word and deed, we will be able to have greater and greater concentration and better meditation. And then we can come in touch with the Divine and attain Realization.

Kundalini

Prayers and other practices purify the mind and stimulate the higher centers of consciousness, known as *Chakras*. There are seven centers or *Chakras*. They are located at the various parts of the *Sushumna nadi* that passes through the center of the spinal cord from the basal plexus to the crown of the head.

The Chakras consist of (1) *Muladhara Chakra* located at the base of the spinal column between the genitals and the anus, (2) *Svadhisthana Chakra* located at the genital area between the navel and sex organ, (3) *Manipura Chakra* located at the navel area, (4) *Anahata Chakra* located between the breasts in the region of the heart, (5) *Vishuddha Chakra* located at the hollow of the throat, (6) *Ajna Chakra* located at the space between the eyebrows, and (7) *Sahasrara Chakra* located at the crown of the head.

The dormant Cosmic energy, known as Kundalini, lies at the bottom of the spinal column like a coiled-up serpent. When the mind is purified by prayer and other spiritual practices, Kundalini energy awakens and passes upward through the Sushumna and lights up the various Chakras. When it reaches *Sahasrara Chakra* at the crown of the head, the Yogi attains the highest state of Consciousness and Enlightenment.

The two lowest centers–*Muladhara* and *Svadhisthana Chakras*–represent the most primitive states of consciousness. When the energy is focused in these centers, the individual remains at the stage of animal tendencies. The third center, *Manipura Chakra*, represents the worldly level of consciousness. The fourth center, *Anahata Chakra,* is the middle one. When the Yogi comes to this stage of unfoldment, he becomes spiritual. He manifests selfless love, friendship, compassion and peace. The fifth center, *Vishuddha Chakra,* represents the state of purity. One who functions at this level exhibits goodness, purity, creativity, intuition and wisdom. The sixth center, *Ajna Chakra,* is the stage of absolute knowledge. The Yogi enjoys absolute freedom and bliss in this state of unfoldment. The final center, *Sahasrara Chakra*, is the seat of Pure Consciousness, the abode of the Lord. The Yogi who operates at this level becomes one with God, the ultimate Reality.

Healing Power of Prayers

Prayer to God with firm faith is a powerful healer. Research studies empirically confirm the great power of prayers and religious faith. Dr, Harold Koenig of Duke University School of Medicine is a pioneer in the scientific study of faith's healing potential. His research team has studied thousands of Americans since 1984. Their studies confirm that prayers promote overall good health and also aid in recovery from serious illness.

Religious patients believe that God is interested in them. This faith serves as a safeguard against psychological isolation.

A Dartmouth Medical School study, and studies made in Israel found that the death rate among religious heart patients was lower by 60% and 40% respectively. Dr. Herbert Benson of Harvard Medical School has found that regular prayer brings a relaxed state and reduces the impact of stress hormones, lowers blood pressure and even slows brain waves, all without drugs or surgery.

The world famous studies of Dr, Dean Ornish, a devotee of Sri Swami Satchidananda, have shown that the practice of Hatha Yoga, a fat-free vegetarian diet, and prayerful meditation cures heart disease without medication. In his recent book, *Love and Survival*, Ornish asserts that achieving emotional tranquility through spiritual practices like prayer could be the "ultimate healing experience."

REFERENCE

1. Eastern and Western Disciples, *Life of Swami Vivekananda*, Calcutta: Advaita Ashrama, 1974, p. 328.

24

Cutivation of Bhakti: Positive Practices-III

In continuation of the previous Chapter, the Positive Practices of 'Meditation' and '*Japa*' are discussed in this chapter.

MEDITATION

Meaning

Meditation is the primary technique of the eight-stage Raja Yoga. It means an uninterrupted focus of the mind on the chosen object of concentration. Meditation is not just an ordinary concentration. It is a special kind of concentration, resulting from spiritual discipline and moral virtues. It is focused on a higher center of consciousness – heart center or any other higher center. The object to be chosen for concentration can be a mystic *mantra, e.g., Om*, the name or form of a Deity, or any other sublime object. Regular and intensive practice of meditation ultimately leads the practitioner to the goal of Self-realization. (For details, consult the Manual on Raja Yoga: Swami Satchidananda, *Raja Yoga: Yoga of Meditation*, Ed. By O.R. Krishnaswami)

Meditation in Bhakti Yoga

Meditation is also a practice in Bhakti Yoga. But it is not a meditation on any object of one's choice. It is, on the other hand, an exclusive meditation on God with form. A devotee focuses his mind on the chosen Personal Deity (*Ishta Devata*) such as Lord

Rama, Lord Krishna, Lord Shiva, Lord Buddha, Jesus Christ or any other form of God. Narada says, "the devotee's whole soul goes toward God, i.e., thought of God to the exclusion of all others. (*N.B.S.*, 9).

Further, the essential characteristics of Bhakti are dedication of all actions to God and extreme anguish if He were to be forgotten (*N.B.S.*, 19). In Bhakti, not only the mind is given to God, but also words and deeds are for Him. Even when the body is engaged in any activity, mind is to be focused on God. Should the mind drift away a bit, it is brought back to the thought of God. This corresponds to the direction given by Lord Krishna in the *Bhagavad Gita*: "Whenever the restless mind wanders away, restrain it and bring it back to the Self." (6. 26). "Supreme Devotion is brought about by unceasing remembrance of God." (*N.B.S.*, 36). Again, God is to be remembered always by the devotee free from all cares and worries, in every aspect of his life. (*N.B.S.*, 79). So Bhakti Yoga is exclusive meditation on God alone.

Concentration

Leading a virtuous life in itself is not sufficient for God-realization. Concentration of mind is absolutely necessary. A good virtuous life only prepares the mind as a fit instrument for concentration and meditation. God is within you. He is seated in the Lotus of your heart. Seek Him through Meditation with a pure mind. Maya or Cosmic Illusion causes havoc through the mind. Conquer Maya through meditation and devotion. (*See* "Overcoming Maya" in chapter 20. Cultivation of Bhakti: Negative Practices–I.) Meditation is an inward journey in the realm of mind. Its aim is God-realization. It is a way of going from the known to the unknown. It gives a direct vision of God wthin the temple of the body. You need not have to search, roam or wander in pursuit of God; you find Him within your heart. Feel the Divine Presence in the lotus of your spiritual heart located in the middle of the chest where you feel the reactions of your emotions. "It is in the heart center God's special power is manifest," says Sri Ramakrishna.

(The Gospel of Sri Ramakrishna). Mahanarayana Upanishad declares:

> "There is a pure, subtle, sinless seat, lotus-like in form,
> Situated in the middle of the body.
> Within this lies the subtle *akasa* (space).
> Practice meditation on this space." (12. 16)

Kabir said,

> Remove the veil (ego)
> You will find your Beloved within.
> In every heart the Lord dwells.
> Therefore, speak no bitter words.
> The One who listens within you
> Also listens within everyone else.

Uddhava, a great devotee and friend of Sri Krishna asked Sri Krishna, "In what form and with what rites should a devotee think of You? Please tell me in detail."

Sri Krishna said,

> "Seated comfortably in a seat neither high nor low, keeping the hands near the body unmoved, control the eyes from wandering outwards. Control the breath by taking it through one nostril and letting it out through the other, and *vice versa*. Control the senses. Repeat Om continuously and with deep devotion both while inhaling and exhaling.
>
> "The above practice daily at three periods (morning, noon and sunset) will enable the jiva to get perfect control of breath within a month. Imagine in the heart the lotus flower with its petals as the seat, and fancy the sun, moon and fire to be the three lights, at their proper places. Concentrate your mind and imagine, "My present form is seated in the flower, with a calm, dignified smiling face."
>
> "Gaze at the whole form till you complete even the details, and then fix your mind on the same without thinking of anything else. By thus looking at the form and fixing the mind without wandering elsewhere, your face will gradually wear an ecstatic look. When thus the mind, oblivious of everything else, gets fixed in concentration on My form,

you become completely merged in Me just as a ball of fire gets merged in a big bonfire. By constant practice of this *Samadhi* or ecstasy you very soon get rid of all delusion caused by diversity, and attain bliss." (*Bhagavatam*)

Practice meditation regularly and without interruption. Surrender or offer your thoughts to the fire of meditation. It means turning your awareness away from your habitual thought waves, and fixing it on the form of God. The thought waves arising out of egoistic desires resist such surrender. Edward Carpenter, a nineteenth century mystic, describes clearly the importance of standing firm in this struggle:

When we inhibit thought and persevere, we come at last to a region of Consciousness behind thought and a realization of an altogether vaster Self than that to which we are accustomed. We leave our ordinary sense of "who we are" and wake up to find that the "I," one's real, most intimate Self, pervades the universe.

When the mind is active, Patanjali says in the *Yoga Sutras*, we identify with our thoughts. When the mind is still, we rest in our true identity, the inner Self or God. The goal of meditation is to surrender to this true Self. Without meditation our minds constantly churn out melodramas of our egocentric life. Once we dissolve our false identification with our thoughts through meditation, we become free of the melodrama.

The profound surrender that occurs in meditation dissolves not only the temporary structure the mind creates, but also the dualism that is built into the mind's way of knowing. By its very nature, the mind divides everything into knower and known, and so we cannot ever *know* our True nature; it is impossible to conceive of non-dual identity. Yet somehow the belief that we are different from what we perceive must go. The Truth is one great Consciousness or all-pervading God. We must know that ourselves to be that God. This is the essence of the surrender that occurs in meditation.

We must give ourselves totally to the ocean of Divine Consciousness. In practical terms this means we must surrender to the Chosen God's form or *mantra* completely. We must identify with It.

Distraction

We may become extremely frustrated when frequent thoughts distract our meditation. The thoughts arise because of our attachments to many external things. Those attachments will not let us go deep in meditation.

When we sit for meditation, thoughts of past impressions stored in the subconscious, thoughts of current problems and imaginations about the future surface. This is why Patanjali says that to attain meditation one must have both steady practice and detachment (*vairagya*).

Control of Distraction

By regular practice and cultivation of moral virtues, we may succeed in minimizing the disturbances of the mind. Through meditation we give a positive theme of the Divine Name that we repeat and the Divine form that we visualize. These serve to focus the mind and hold our attention within. We must think of God with love and devotion in our heart. The Name, the Divine Form and the Love for God keep the mind concentrated within. When our interest in God becomes greater than our interest in outside things, the visualization of God's Form becomes more and more clear and real to us. The mind begins to dwell on the blissful Divine Form and on the Divine glories and attributes. Then comes the time when one feels the Divine Presence. At this stage true devotees are blessed with the spiritual vision of God, as it has been found in the lives of great devotees. Tukaram, an Indian saint of seventeenth century, says,

> "By meditating on the Lord this mind and body become transformed.
> How can I explain the fact that my I-ness has become the Lord Himself?
> When the mind embraces the supreme Consciousness
> The entire universe becomes the Lord. Tuka says,
> How can I describe how my whole being is immersed in the Lord?"

Stages in Meditation

The ultimate purpose of mediation is to merge our consciousness in Cosmic Consciousness and thus to attain the state of Super-consciousness. This is beyond the reach of anyone in the initial stage. So a graduated course is followed, proceeding stage by stage.

- The first stage is to meditate on the blissful form of the Chosen Deity.
- The next stage is to meditate upon the infinite qualities of God such as infinite Purity, Knowledge, Love and Bliss.
- The third and the final stage is to meditate on the all-pervading Consciousness of which the Holy form of Deity, the Guru, the disciple and the universe are different manifestations. This infinite background should never be forgotten even during the previous stages of meditation. It is good to meditate on the infinite Spirit, merging all forms into It, and try to be established in the Supreme Consciousness through repeated practice.

Requirements of Meditation

The Yoga of Meditation requires a disciplined life and cultivation of ethical and moral values. (For details, *see* chapter 22. Cultivation of Bhakti: Positive Practices–I.) Meditation cannot be achieved without integrating and perfecting one's life. Therefore, an aspirant has to change his way of life and mental outlook according to the requirements of meditation, which are briefly described below.

1. The body must be healthy and fit for the practice of meditation. Sound health is essential for maintaining sound mind. Health can be maintained by disciplined life-style, sattvic food, and by the practice of Hatha Yoga asanas and Pranayama (breathing exercises).
2. Sattvic foods are the most desirable ones for spiritual life. They include fruits, vegetables, nuts, seeds, pure milk and grains. Meat, refined foods like white sugar, preservative-added foods and alcohol should be avoided.
3. Moderation in diet, work, rest and sleep is essential for Yogic life.
4. Peaceful living is essential. A prayerful attitude to work is

most helpful. You must always be aware of the Divine Presence in every situation. This helps to prevent distracting thoughts and useless desires entering into your mind.

5. Positive outlook is required. You have to be loving and become a channel for carrying the message of love and friendship to all beings.

6. You have to cultivate positive and benevolent attitudes toward people in order to maintain mental peace.

7. Holy association (*Satsang*) is conducive to meditation.

8. A clear understanding of the basic purpose of the spiritual practice is essential. Its purpose is to realize God.

9. The cultivation of surrender and devotion to God helps to overcome your ego.

10. Dispassion toward sense objects, and discrimination between the real and the unreal play an important part in self-discipline.

Preparation for Meditation

Meditation requires adequate preparation as described below.

1. Select a clean, quiet and secluded place or room for practicing meditation. Keep it holy like a place of worship.

2. Select a suitable time. Early morning, noon and evening (sunset) hours are the best time for practice.

3. Keep the stomach, bowels and bladder empty. Take a bath to feel fresh for meditation.

4. Do some Hatha Yoga asanas and Pranayama exercises in order to keep the body and mind fit for meditation.

5. Sit in a comfortable cross-legged position, facing the East or North. This will create a positive effect. Keep the head, neck and spine erect in one straight line.

6. Begin your practice with a chant or prayer.

7. Assume an attitude of detachment. Gently close the eyes, and withdraw the mind from sense objects.

8. Focus the mind on the object of Meditation – a *mantra* or the Name and Form of the Chosen Personal God. Start repeating the *mantra* or Divine Name mentally.

9. In the beginning meditate for about ten or fifteen minutes. After gaining some experience gradually increase the duration of the time period.

10. Whenever the mental attention drifts away from the object of meditation, bring the mind back to the point again and again. By regular practice you will be able to retain the focus of the mind on the object of meditation for a longer time.

11. Practice regularly with perseverance and devotion. You will eventually reach the state of God-realization. Of course, you need the guidance and grace of the Guru and the Grace of God.

JAPA

Meaning

Japa means repetition of the *mantra* or Divine Name imparted by the Guru. It can be practiced by all irrespective of age, sex, caste and stages of life, at all times and in all places. It is to be done in a spirit of mystic worship. When this spirit is lost, the repetition becomes mechanical. The attitude of Love and adoration to the Divine must be there to make the *Japa* an effective spiritual discipline. When it is done with devotion, our mind, body and soul respond to it. This is the secret of attaining success in spiritual life through *Japa*. Choose whichever version of the holy Name that appeals to you. Once you have chosen a holy Name stick to it and do not change. Otherwise you will be like a person digging little holes in many places without going deep enough to find water. *Japa* implies also dwelling on the Divine attributes. This means that along with the repetition of the Name, think about the Divine virtues like love, compassion, purity, etc., or visualize a holy Form. Visualization and *Japa* go hand in hand.

We need a definite holy Form, so long as we consider our forms to be real. But we must know the connection between the Form and Formless. Form is only a manifestation of the Formless. The holy Personality is a manifestation of the Formless Brahman or Supreme Consciousness.

Just as the form is a symbol of Divinity, the eternal sound Om is also a symbol of Divinity. We take the help of both the symbols to call up the Divine Consciousness.

Japa is an all-inclusive and all-sufficient practice by which you can be ever in tune with God and finally merge yourself in Him. *Japa* is thinking of God constantly to the exclusion of everything else. God-thought keeps away every other thought. This thought also ultimately disappears and you realize your identity with God.

Importance

The *Yogacudamani Upanishad* (87, 88) emphasizes the value of constant *Japa* for self-purification and Self-realization. *Mahabharatam* declares that *Japa* is the best of all spiritual practices. The *Bhagavatam* observes: "That which one obtains through meditation in Satya Yuga, through sacrifice in Treta Yuga, through worship in Dvapara Yuga, may be attained in Kali Yuga by reciting the names of the Lord." (12. 3. 52). *Manusmrti* says, "The seeker after Truth reaches the highest goal by *Japa* alone. (2. 87)

Buddhism, Christianity and Islam also have the same consideration for the Divine Name. "Let them also love Thy name be joyful in Thee." (*The Psalms*, 5.11). *Japa* is superior to all rites. Mental repetition is the most effective.

Japa in the World Religions

The repetition of the Holy Name is a practice found in all major religious traditions. In Hinduism, great importance is given to *Japa*. The Saiva and Vaisnava Saints of Southern India and the great Saints of Northern India placed great emphasis on the constant repetition of the Divine Name. The observations of important scriptures about *Japa* have been mentioned under "Importance" above.

In Buddhism, the main emphasis is on ethical living and meditation. But some schools of the Mahayana Buddhism prescribe the repetition of the Divine Name as a means of Enlightenment.

Many Christians simply repeat "Jesus, Jesus." The Desert Fathers repeated the Prayer of Jesus: "Lord Jesus Christ, have mercy upon me." Some Eastern Orthodox Christians use this Prayer even today. Catholics use *"Hail Mary"* or *"Ave Maria."* In the Greek Orthodox Church, great importance is given to the repetition of a prayer that resembles the Hindu *Japa*. The Greek saints of the Middle Ages perfected a technique of repetition of the Jesus Prayer, "Lord Jesus Christ, Have mercy on me."

In Islam, the Sufi mystics have employed the repetition of "Allah" as a means of getting spiritual Illumination.

Guru Nanak and his followers attached great importance to the Divine Name. Caitanya Mahaprabhu made *Japa* popular among all sections of people in Bengal.

Sri Ramakrishna said, "When you chant God's name with single-minded devotion you can see God's form and realize Him.

By repeating God's name you become absorbed in Him and finally realize Him." The Holy Mother used to practice *Japa* during her sadhana period – about a lakh of *Japa* a day. Through *Japa* and meditation, she would lose body consciousness and rise to the heights of Super-conscious experience. In her teachings she emphasized the importance of *Japa*: "You will realize everything by the repetition of God's name. The mind will be steadied if one repeats the name of God fifteen or twenty thousand times a day."

Purpose

The purpose of *Japa* is to awaken the forgotten spiritual tendency in the aspirant. Through regular practice of *Japa*, the energy stored as impressions in the mind is sublimated and directed into higher channels. The sublimation purifies the mind and awakens intuition. The repetition of the Divine Name along with the contemplation of the meaning or the Ideal it stands for gradually removes the obstacles and makes the mind introspective. *Japa* puts the mind in tune with higher Cosmic vibrations. This calms, elevates and concentrates the mind. Constant practice of *Japa* and meditation on the Divine Spirit finally enables the devotee to experience the presence of God in himself and in others.

Types of Japa

The methods of recitation are described in *Smrtis*.[1] *Japa* can be performed in three ways–loudly, uttering to oneself, and repeating merely in mind without the least movement of vocal organs. The mental *Japa* is the most efficacious of all; and it can be practiced even when the hands are employed in activities.

Conditions

The Requirements of Meditation and Preparation for Meditation listed under Meditation, above, are applicable to *Japa* as well. The scriptures specify certain conditions for ensuring success in *Japa*.

1. The *Prapancasara* says that you should have a light stomach, and should have had sound sleep, before you sit for Meditation or *Japa* on a proper seat in an equable spot, with closed eyes and turning toward the East.
2. You should sit straight with the palms of the hand placed on the lap, right over the left, well stretched.
3. You should keep your mind alert and calm. Purity, silence, reflection on the meaning of *mantra*, freedom from distraction, and absence of indifference contribute to success in *Japa*.
4. Sitting posture is very necessary. If *Japa* is done while standing or walking, attention will be distracted by the strain of the muscles, and if it is done lying down you may easily go to sleep. So a sitting posture is better for concentration.
5. You must have a Chosen Deity, a particular *mantra* and a definite center of consciousness such as the heart center or space between the eyebrows. You must hold on to that center while doing *Japa*.
6. Practice *Japa* and meditation regularly. Do not miss even one day.
7. In the beginning you may face a lot of resistance from the body and the mind. But be persistent in your practice. You

will gradually overcome that resistance and bring the body and mind under control.

8. Contemplating in the mind, repeat the *mantra* vigilantly.

9. In order to get perfection in *Japa* you should know the meaning and Deity ensouled in the *mantra*. Repetition must be continued until perfection is reached.

10. The mind can grasp only attributes, and God is the repository of all perfection. Therefore you have to meditate upon the blessed attributes of God, as Narada advocates in the *Bhakti Sutras*.

11. Faith in the power of *mantra* or Divine Name is most essential.

12. As Swami Ramdas said, you must be aware, while doing *Japa*, that you are repeating the name of One who is within yourself. Without it, the repetition becomes merely mechanical and does not help you in any way. When you tune yourself with the Name, you tune yourself with God. You must keep this central fact in your mind when doing *Japa*. Then the practice will gradually make you aware of the Divine existence within yourself. This idea grows into an experience when you actually feel God's presence. As you become aware of Him within you, you become aware of Him also without you. You feel his presence everywhere.[2]

13. Do not say that you have no time for doing *Japa* and meditation. If you do not find time for your daily spiritual practice, you can never progress. Avoid all unnecessary waste of time in futile thinking, gossiping, meaningless activities, etc., then you will get plenty of time for your spiritual practice. Devote more and more time to your *Japa*, meditation and spiritual studies. Then you will advance in spiritual life and reach the goal of life.

Japa with Mala (Rosary)

Japa with Mala helps you to keep the mind alert and to fix it constantly on the *mantra*. *Japa-mala* (rosary) contains one hundred and eight beads made out of the roots of the tulsi plant or of glass or sandal wood, according to the Hindu tradition. One hundred and eight is an auspicious number. One represents the process of evolution towards non-duality. Eight represents Maya, the Divine illusion through which the soul evolves, because the multiples of 8 (16, 24, 32, etc.) are gradually lesser than 8 when each component number is added together (1plus 6 equals 7, 2 plus 4 equals 6, 3 plus 2 equals 5, etc.). Though apparently more in value, the essence is less. 0 represents the world which has value because of 1 being added to it – because of evolving souls. 108 represents the Absolute, *Brahman,* ever constant in all changes and modifications, the goal of all practices. The component numbers of 108 add up to 9. And all the multiples of 9, when their component numbers are added, total 9. Therefore 9 represents the highest number, and the highest attainment.[3]

Where the two ends of the rosary are brought together into a knot there is a bigger bead. This is called the Meru and considered to be the seat of the *Mantra*-Deity. In counting the beads, you should start clockwise from the first bead from the Meru. When you complete the circle and reach the Meru again, turn the rosary inside out and start clockwise, keep the rosary on the ring finger, and use the thumb and the middle finger to count and slide.

Power of Japa

The *mantra* or Divine Name has tremendous power. It will remove obstacles and awaken the spiritual Consciousness. Through proper *Japa* apparently dead sounds become living and acquire tremendous power. This power can be realized only if you practice spiritual disciplines systematically and lead a pure life. Through proper repetition of the *mantra*, you can attain the highest illumination and freedom. Through perfection of *Japa* and meditation, the Supreme Spirit manifests. Then your soul is united with the Supreme Spirit or God.

God-realization is possible only when the mind is controlled and purified. Constant *Japa* arrests the restless nature of the mind.

It conserves all the physical and mental energies for removing the veil from the indwelling God. The *mantra* or Divine Name is an unfailing key that unlocks the gates of the heart, permitting an outflow of immortal love, wisdom and power. "God's all-powerful Name," says Swami Ramdas, "takes the aspiring soul to the highest summit of Truth. What is required first is an absolute faith in the greatness and potency of the Name, which comes only by the Grace of the Lord. When the Name becomes the sole mainstay and refuge of the aspirant who thirsts for the highest goal of life – God-realization – he or she marches towards the goal not only in rapid strides but also with a heart filled with courage and cheerfulness." (*The Divine Life*, p. 250)

Japa and Control of Mind

Of all the disciplines for controlling the restless mind, there is none so easy and efficacious as the repetition of God's Name. But the difficulty with some is that they cannot repeat the Name continuously although they desire to do so. The reason for this is that their love for the Name is not greater than their love for the perishable objects of the world. As we think so we become. So if our mind is fired with an intense love for the Divine Name, this love will automatically enable us to repeat the Name constantly. When thus the mind is inebriated with the love of God and filled with the music of His Name, we will experience the Divine ecstasy. When we are in this joy, the vision of God, says Swami Ramdas, will flash out. Thus *Japa* is the easiest and best method by which we can purify ourselves preparatory to God-experience. Swami Ramdas affirms that no other spiritual practice could so easily grant us purity – absolute freedom from lust, greed and wrath. There is no other easier way to quell the waves of the mind and free it from all kinds of desires.

The repetition of the Divine Name subdues the waves of the mind and gives us strength, steadiness and peace. If we have a real longing to realize God, His Name will cast a spell on us and prevent our mind from wandering. The Name stills the mind and floods our entire being with bliss. Edward Carpenter, while in

Ceylon (now Sri Lanka), met a Jnani. When they went out for a walk, Edward Carpenter asked the Jnani, "What is your state of mind?" He replied in Tamil, "*Anandam, eppodum anandam* – Bliss, always bliss."

There is no greater victory in our life than the victory over the mind. He who has controlled the gusts of passion that arise within him and the violent actions that proceed therefrom is the real hero. It is the constant *Japa* that makes this feat possible in the easiest way.

Swami Ramdas (1884 – 1963), a great Saint of South India, reached the heights of God-realization only by taking to the chanting of the *mantra 'Om Sri Ram, Jai Ram Jai Ram.'* This constant chanting of the *Mantra* enabled him to remember and feel the presence of God within him and everywhere around him, and to realize that the entire universe was the form of God. This naturally brought him to the absolute state of surrender, i.e., accepting whatever happened as happened by the Will of God. He describes the greatness of Divine Name as under:

> "It is said that God's Name is greater than God. Ramdas knows from experience that no other spiritual practice can so easily grant you purity – absolute freedom from lust, greed and wrath – and make your mind just like that of an innocent child. When you have the Name in your mind, there is light, purity, peace and joy. It is not that you get happiness some time after repeating the Name. The moment you repeat it with love for God, the moment you have it on your tongue you enjoy its sweetness. The simple Name of God has this wonderful potency, because the name itself is God." (*The Divine Life*, p.253)

REFERENCES

1. The law books, subsidiary to the *Vedas*, guiding the daily life and conduct of the Hindus.
2. Swami Ramdas, *The Divine Life*, Anandashram, 1991, p. 251
3. Swami Jyotirmayananda, *Mantra, Kirtana, Yantra and Tantra*, Miami: International Yoga Society, 1975, p.25

25

Cultivation of Bhakti: Positive Practices-IV

In continuation of the previous chapter, the Practice of 'Self-surrender' is discussed in this chapter.

SELF-SURRENDER

Meaning

Self-surrender means offering our body, mind, will and soul at the feet of God, without any sense of ego. It is complete dedication of ourselves to the Supreme Spirit. Then as Tukaram sings, the self (ego) within us is dead and God is enthroned in its stead, and there is no longer any "me, my or mine." Without complete detachment to the objects of the world and our desires and prejudices self-surrender is impossible. Self-surrender is a state in which the thought of God runs in the mind in a ceaseless and continuous stream. So constant remembrance and meditation of God automatically lead to self-surrender. Self-surrender involves a total change in our attitude towards ourselves, towards the world, and towards God. Is there anything that we can claim as 'ours'? There is only one thing; that is God. Everything else – human beings, other living beings, objects of the world, the entire universe – all is God's manifestation. So it is foolish to think that we are separate entities and that possessions belong to us.

Perfect self-surrender is the culmination of spiritual life. It can be practiced only by those who have greatly advanced in spiritual life.

It is the final step in the Path of Bhakti (Devotion). The devotee makes God the center of his life. He offers himself – body, mind and soul – to God. When the ego dies God reveals Himself and blesses the Devotee with God-realization. What is achieved by the strenuous practice of the eightfold Raja Yoga can be achieved by self-surrender. Patanjali, in his *Yoga Sutras*, states that by devotion and self-surrender alone one may attain *Samadhi*. (1. 23).

In Sanskrit the word for surrender is *sharana* which means 'taking refuge.' It implies not loss of one's own will, but linking oneself to the higher will. As a river merges into the ocean, a seeker merges his small will into the greater Power, which is the doer of all things. By taking refuge in God, by opening himself to the Divine Will, he sees that there are not millions of wills in this universe, not millions of separate minds, but only one Will, only one Mind. Like the leaves of a huge tree, we are all part of that One force. God is not separate from us. The Will of God is really the will of our own higher Self. When we recognize this force and align ourselves with It, then our lives can begin to become integrated, focused and free.

The Divine Will manifests in the form of inner intuition. It is the inner voice that is often drowned out by other voices in the mind. But we have to learn to hear it and to obey it with the grace of the spiritual Master, Guru and the Grace of God. When we become aware that we are not the doers, we become a real instrument of God, and God acts through us. "When you are acting in accordance with God's will, the very power of God takes care of everything completely. It is both the path and the torch that reveals the path," says Gurumayi Chidvilasananda.

Stages in Self-surrender

There are six stages in the ascending order of self-surrender:

1. A devotee aspires to experience the sweetness of self-surrender to God.
2. He turns away from all that is contrary to the development of devotion. He avoids evil association, and gives up anger, hatred, jealousy, lust, greed, pride, ambition and all emotions that are obstacles to the unfolding of Divine sentiment.

3. The devotee develops unshakable faith in God and His plan through creation.

4. He sticks to the Path of Devotion. Just as a miser fixes his mind on his wealth, so the devotee centers his mind on God at all times.

5. He develops a deep relationship with God. A stream of Divine feeling flows on in the depth of his heart.

6. Finally he experiences the joy of self-effacement. All that was his—including his body, mind, intellect and soul—has become the possession of the Divine Self. This is the highest form of consecration. The effacement of the ego-sense is the culmination of renunciation. The devotee has no longer any sense of doer. He has dedicated his life to the Lord. He experiences inaction in action. He is not bound by the actions he performs. He lives in the spirit of the following ancient prayer:

> "Whatever I do with my body, speech, mind, intellect
> and senses led by the Gunas of Nature
> I hereby surrender completely to the Supreme Lord."

Self-surrender becomes perfect only after God-realization.

The Purpose of Self-surrender

We undergo sufferings because we are ignorant of our true being. The ignorance resides in our sense of ego, our belief that we are individual beings. The ego veils our eyes, as it were, and causes us to dwell in ignorance. We are the Atman, the Spirit within ourselves. When we forget that we are the Atman, and identify ourselves with the body, mind and senses, then the sense of ego arises. With the birth of this ego-sense, our transcendental nature is forgotten, and we live on the sense-plane and become subject to the Law of Karma and rebirth. In order to fulfill our desires for sense-objects, we are driven to do all sorts of actions. And our actions, in their turn, involve us in the bondage of Karma. We get attached to what is pleasant, and dislike what is unpleasant. Out of this attachment and aversion there grows a clinging to life. The ego clings to its ego-life, its sense of individuality.

Therefore, to find the real life, the life of our true nature, we must transcend the ego and realize our Self. Then only can we regain our inherent peace and happiness. The ego is the only barrier to this knowledge of Self. Sri Ramakrishna used to say that when the ego dies all troubles cease.

Self-surrender is the best means for killing the ego. It is by the total self-surrender that we can dissolve the sense of ego in the consciousness of God. We must love God with all our hearts, all our souls and all our minds, and become absorbed in Him, and forget ourselves in the consciousness of God.

Approaches to Self-surrender

There are two approaches to self-surrender. Some devotees feel, "I have to surrender. I have to depend on God, but still I have to hold onto Him." Have you ever seen the mother monkey jumping around with the baby monkey? The baby monkey clutches the stomach of the mother and holds it with the forepaws and the legs. It holds her entire body, hanging underneath. The mother monkey does not even worry about the baby. She just jumps from branch to branch. It is baby's duty to cling to the mother. Whenever the baby wants to go somewhere, all it has to do is to take hold of the mother, and then the mother jumps. When the mother jumps from branch to branch, which has the fear of falling down? The baby. It is depending upon its mother, no doubt, but it still depends on its own strength. So also this type of devotee feels, "God, I won't leave You. I'm holding on tight."

There is another animal that also depends on its mother for movement – the kitten. Whenever the kitten wants to move, all it has to do is meow. The mother comes, picks it up by the back of the neck, and jumps here and there. When the mother jumps with the baby, who has the fear of dropping the baby? The mother. Both the baby monkey and the kitten depend upon their mothers for their movement. But in the first case, the anxiety is in the heart of the baby monkey, whereas with the cat, the responsibility rests with the mother. The kitten is totally free. Some devotees are of this type. Their surrender is absolute.

So there are two different views on surrendering. One view is: "You must be like a little monkey and hold on , and then God will take you." Another view is: "Even to hold on, where would I get the strength if God didn't give it to me? So, why should I worry? I am God's child."

It is very difficult to have that kind of total surrender to God, because the ego holds its supremacy. It will not give up that easily; there will be a big fight. You have to prove your capacity.

If you totally surrender yourself into the hands of God, you will feel, "Lord, I didn't plan to come here. You sent me here. You have a purpose. I don't know what that purpose is. If I could ask You for anything, it would be this: 'Let me always remember that You are working through me. You are solely responsible for everything. I am nothing. You are everything. Please let me not forget this.'"

Free Will

You might ask, "Then what is free will?" Free will means you are free to take responsibility into your own hands or to give it into the hands of God. If you take the responsibility yourself, after going through all the problems, difficulties and turmoil, you will finally say, "I am tired, Lord. I'm not going to try to handle this anymore. I know I cannot. So let me put the reins in Your hands and take the back seat. You drive."

That happens when your ego has been completely cleansed. Then, even though it might appear to others that you are doing something, you will know that you are not doing it. You are being made to do it. God's power will make you think, make you do. Your job is to simply allow it to happen. Then your life will be ever peaceful.

Separateness

One who wants to retain his separateness as an individual cannot make a real self-surrender; his so-called surrender with that reservation does not destroy his ego-sense.

Once a young man came to Ramana Maharishi,[1] in a disordered state of mind. He believed he had had a vision of God, in which he was promised great things if he surrendered himself. He said he had done so, but that God failed to carry out His promise. He demanded the Maharishi: "Show me God and I shall chop off His head, or let Him chop off my head." Maharishi asked someone to read from a Tamil commentary on his own writings, and then made this remark: "If the surrender is real, then who is there remaining and able to question God's doings?" The young man's eyes were opened; he acknowledged that it was his own mistake and went away pacified. Self-surrender must be without any reservations and without conditions; there is no room in it for bargaining.

You achieve perfect surrender only when you give everything – body, mind, senses and soul – completely to God without holding back anything. Your surrender should not be for any selfish motive, nor should you make a show of surrender. It should be genuine, real surrender with all your heart, with all your being. The Lord takes the responsibility for the redemption of one who takes total refuge in Him.

Tests

God wants proof that you are totally surrendering yourself to Him. Tests will come. If you fail the tests, then your surrender is not complete. If your surrender is complete, no matter what test comes, you will pass it. Once you pass the tests, you do not have to worry about anything that happens in your life, because you are not responsible for your actions. Whatever comes, comes from God. Whatever goes is taken away by God. It is not that easy in the beginning because the ego will not allow you to surrender in this way. Until you become the master of the ego and give yourself into the hands of God, the ego will create doubts in you.

The self-arrogating little ego persists and resists again and again. It clings leech-like to its old habits, cravings and desires. It resists surrender. That is the reason why Lord Krishna says, "Flee unto Me for shelter with all thy being, O Arjuna. By My Grace thou shalt obtain supreme peace and the eternal abode." (*Bhagavad Gita*, 18. 62). The ego, the mind, the intellect and the soul should be placed at the feet of the Lord. Mira did this and so she obtained Lord

Krishna's Grace and became one with Him. (For her brief biography, *see* chapter 33. World Teachers and Great Devotees.)

Total Surrender

One attains this highest spiritual stage only after prolonged struggle and intense spiritual practice. All forms of spiritual striving make the mind pure and fit for this total self-surrender.

Total self-surrender can come only when our 'wings' are dead-tired like those of the bird in Sri Ramakrishna's parable:

A bird was sitting on the mast of a ship. It did not know that the ship was moving. When it suddenly realized that, it started flying towards East, West, North and South but found only limitless sea everywhere. Finally when it wings were dead-tired, it came back and sat on the mast, allowing itself to be carried by the ship.

In the same way, only a person who has put forth such maximum self-effort in the form of intense spiritual practice with great perseverance can know what true self-surrender is. It is not enough to say simply "All is Thine; Thy will be done." Self-surrender is an attitude of mind. One acquires it through the awareness that one's soul is a part of the Cosmic Spirit or God and that one's body and mind are instruments of that Spirit or God.

According to Ramana Maharishi, self-surrender is surrendering oneself to the Source of oneself. One's Source is within oneself. To that Source the surrender should be made, i.e., one should seek that Source and by the very force of that search merge into It. If the ego becomes merged into its Source, the Self, then there is no ego, no sense of individuality, no individual soul; that is, the seeker becomes One with the Source. That being the case, where is surrendering? Who is to surrender and to whom? And what is there to be surrendered? The loss of individuality. Even this does not really exist. It is only an imaginary belief. The Vaishnava Saint Nammazhvar of Southern India sang, "Not knowing the truth of myself I was deluded by the ideas of 'I' and 'mine;' but when I came to know myself, I know also that Thou art both *I* and *mine*." The verse of the *Bhagavad Gita*, "I am the Self residing in the hearts of all" makes it clear that God Himself is the real Self in all.

Self-surrender--not a Gift

Self-surrender is not a gift of oneself to God. "The offering of oneself to God is similar to the offering, to a Ganesha figure made of jaggery, of a portion of jaggery taken from the figure itself; for there is no individual Self apart from God, says Ramana Maharishi. What is really meant by the term 'Self-surrender' is just the recognition by the devotee that he himself has no separate existence. That is why Lord Krishna, in the *Bhagavad Gita*, exhorts us:

> Fix thy mind on Me, be devoted to Me, sacrifice unto Me, bow down to Me;
> having thus united thy whole self with Me, taking Me as the Supreme Goal,
> thou shalt verily come unto Me. (9. 34).

One Consciousness

The *Upanishads*, the *Bhagavad Gita* and other scriptures say: There is only one Consciousness everywhere in this universe, and we are part of It. When through the grace of the Guru, we begin to sense the all-pervasiveness of that great Force we realize that surrender to God's Will is not really a matter of choice. Whether we accept it or not, God's Will is already being done. God will go on being the doer. Without His will, not even a leaf can move.

> The Lord abides in the heart of all beings, O Arjuna,
> causing them to turn round by His power as if
> they were mounted on a machine. (*Gita*)

All creatures, according to their inherent nature, are activated by the Supreme Being. Our past, present and future rest with the Lord. Let us play our part as best as we can as His instrument. The Lord dwells in our heart and guides our destiny. Let us gain freedom from anxiety and tensions and attain peace with His Grace.

Real surrender

A beautiful story in the ancient Epic, the *Mahabharatam*, clearly demonstrates what is real surrender to God:

The Pandavas and the Kauravas were cousins in the Royal family. After the death of the father of the Pandavas, all the children were raised together in one household. The Pandavas were loved by all for their virtue and valor, and this made their cousins jealous. The eldest of the Kauravas, Duryodhana, could not bear the growing wealth and fame of the Pandavas and decided on a plot to dishonor them.

The eldest Pandava, Yudhisthira, was tricked into playing a game of dice in which he lost all that he had, including his wife. Draupadi, his beautiful queen was dragged into the court. Duryodhana decided to put Draupadi to shame by stripping off her sari. By doing this, he would ruin the reputation of the Pandavas.

A sari is six or seven yards of material that is gracefully wrapped around and tucked in with delicate folds. Duryodhana's brother began to pull Draupadi's sari, and though she was clutching it tightly, very soon he succeeded in pulling out one round and then almost a second round of the sari. Draupadi was very frightened; she was calling out to Lord Krishna.

"Krishna, Krishna!" she cried, but her cries went unanswered. In another minute, the entire sari would be pulled off. Her strength could never match her assailant's. Realizing the situation she thought to herself, "This is the final round of my sari; if I lose this also, I will be disgraced." In that moment a great realization dawned, "What am I doing? I cannot take care of myself anymore. Lord Krishna, if you want me to face this disgrace, I will accept it. I totally trust you; my life is in your hands." And with that she let go of the sari and held up her hands crying, "Krishna!"

Draupadi just stood there calling and crying to Lord Krishna while Duryodhana's brother mercilessly pulled the last round of her sari. But as he pulled, the cloth kept coming. After the last round, there was another, and another and still another. He pulled again and again, yet there seemed to be no end to the sari. He was pulling and pulling and pulling. Yards and yards and yards of sari seemed to be coming from somewhere. He was soon exhausted and could pull no more. Draupadi was saved.

Even God cannot come and help you as long as you have faith in your own strength. Complete surrender means to give up totally and depend on God.

Swami Ramdas (AD1884-1963)

Swami Ramdas, founder of Anandashram, Kanhangad, Southern India, was a great Saint. He was a recent example for total surrender to the Lord Ram. His name was fitting, for Ramdas means the servant of Ram.

He saw Ram in every face. Somebody once said to him, "Why don't you go to Benares (a holy place in North India)?"

"Okay. Ram is telling me to go to Benares. Well, how should I go?"

"Go to the railway station and take the train."

He went to the station and when the train came, he got in. The ticket collector asked, "Where is your ticket?"

"Oh, Ram didn't tell me about that."

"Well you can't continue if you don't have a ticket."

"Okay, what do you want me to do?"

"You have to get off at the next station."

"Fine." He got off at the next station. He simply felt, "Ram asked me to get off, so I got off," and that is all. He just sat there on the platform until somebody came and said, "Why are you sitting here?'

"Ram wanted me to sit here."

"How did you come here?"

"Oh, Ram said I should go to Benares on the train, and again Ram came and asked me to get off."

"Oh, that's not the right approach. You are simply listening to every Tom, Dick and Harry."

"No, It's Ram. Ram as Dick, Ram as Harry, the same Ram."

"So who am I then?"

"You are also Ram."

"Okay, then stand up."

"Sit down." He sat. "Stand up." That fellow made him stand up and sit down fifteen times. Still Ramdas felt, "Ram is saying so, all right, I'll do it."

Finally the fellow realized the greatness of this person. He felt terrible and fell at Ramdas' feet saying, "I'm so sorry. Please forgive me."

"Oh, Ram, you shouldn't do that."

"Come on, the next train is bound for Benares. You get into it."

"All right," and he boarded the train once again. It didn't matter who said what to him, it all came from Ram.

REFERENCE

1. A great Sage lived in Tiruvannamalai in Southern India during twentieth century. He attained Super- consciousness through Jnana Yoga.

26

Cultivation of Bhakti Positive Practices-V

In continuation of the Previous Chapter, the Positive Practices of 'Selfless Service,' 'Dedication of Actions,' 'Giving up Fruits of Actions,' and 'Reflecting upon Scriptures' are discussed in this Chapter.

SELFLESS SERVICE

Meaning

Selfless Service means service to human beings and other living beings without any expectations. When such service is rendered in a spirit of service to God, it becomes a form of worship. Service is rendered to the Guru (the Spiritual Master), Sages and Saints, devotees, helpless people, destitute, sick and elderly people. Service is also rendered at holy places, and temples, churches or mosques.

The way of Performing Actions

The way a person performs his work is crucial. In the *Bhagavad Gita*, Sri Krishna makes three distinctions about actions.

1. *Rajas*ic Actions. A passionate person performs actions out of longing for the fulfillment of desires or personal gains with egoism (*Gita*, 18. 24). Such actions are prompted by selfish motives. Even the charities given by a passionate person are motivated by desire for name, fame and publicity. He can never work without the expectation of a reward. He is cruel, greedy and impure. He strives

to obtain gains. He rejoices in success and grieves in failure (*Gita*, 18. 27).

2. *Tamasic Actions*: These are actions undertaken out of delusion without (removed 1 word here) regard to the consequences of loss, injury and one's own ability to complete the work. (*Gita*, 18. 25). A *Tamas*ic person does not discriminate between good and bad. He acts blindly. He is unsteady, vulgar, unbending, malicious and cheating. (*Gita*, 18. 28). In every action he is guided by his own selfish interest, impulses and desires.

3. *Sattvic Actions:* A person of purity enthusiastically performs actions without any expectation of reward, without any thought of their fruits. He is neither elated by success nor grieved by failure. (*Gita*, 18. 26) He does his duty with joy and offers its fruits to the Lord. (*Gita*, 18. 23). *Sattvic* Actions are selfless actions.

Attitude of Server

While rendering service to others, you should guard against the development of subtle forms of pride. Any thought that you are helping others strengthens your sense of egoism and obstructs your spiritual progress. Service for the sake of name and fame is not selfless service. There should be no such personal motive in doing service. You should rather feel grateful to the recipient for giving you an opportunity to enjoy the joy of service, an opportunity to serve God through him. The best way to avoid any development of sense of pride is to consider yourself as the humble instrument of God and the served as God in human form. You must render service with heart and soul whenever you get an opportunity to be of service to others. There is no superior or inferior service. Every work is Divine service. Every work is worship of the Lord. Mahatma Gandhi of India never made any difference between menial service and dignified work. Cleaning of the latrine was the highest Yoga for him.

Do service to others with the feeling that God dwells in all and receives your service as worship. The world is nothing but a manifestation of God. Service is worship of God. It is God who prompts you to action. You are only His instrument. There will be no 'mine-ness,' only 'His-ness.'

Do every work with concentration and devotion. Do not think of success or failure. Perform work without attachment, without the feeling of doing it. Whenever your inner voice commands you to give up the work, you must at once relinquish it. Attachment to work will bind you. Work as selflessly as the clouds that shower rain.

Absence of the sense of doer-ship is the key to selfless service. If you think, "I am doing this, I am the doer," your action loses all its power. "Forget yourself in the service of humanity," says Swami Sivananda.

Do not feel satisfied with working for your own illumination and liberation. Assist others in attaining illumination. This is the greatest of all services. Other services like serving the sick and poor, giving food to the hungry and the like, though necessary, give only temporary benefit to the recipients. But spiritual service, service of Enlightenment gives lasting benefit to the served.

The Effects of Service

Selfless service is the watchword in the spiritual path. It purifies your heart and makes it a fit abode for God to dwell. It fills your heart with Divine virtues. Humility, pure love, sympathy, tolerance and mercy are developed. Egoism, hatred, jealousy and selfishness vanish. The idea of superiority goes away. The sense of separateness is annihilated. You begin to feel the oneness or unity of life. You develop a broad heart with broad, generous views. Eventually you get the knowledge of Self. You realize the 'One-in-all' and 'all-in-One.'

By living for others, you will attain the state of blessedness. The world is yourself. "Therefore love all, serve all," says Swami Sivananda, "be kind to all. Behold the Lord in the poor, the down-trodden, the oppressed and lowly ones."

Service, performed in a spirit of worship of God, tends to liberate you from worldliness. It makes you free from anxieties, disappointment and tension. Instead it gives you joy, peace and satisfaction.

You derive the capacity to serve or to work from God who dwells within you. Feel that God is working through you and feel His presence in all.

DEDICATION OF ALL ACTIONS

Dedicate all your actions to God. You should also direct all your passions such as lust, anger, pride and so forth to God too. (*N.B.S.*, 67). One of the religious ceremonies in the Hindu tradition, is a fire ceremony called *Homa*. It symbolizes dedication of actions. A fire is ignited. God is invoked into the fire. Oblations are offered and prayers are chanted. At the end of the ceremony all actions and their effects, good or bad, are dedicated to God with the following prayer:

> "I, an embodied being, endowed with intellect, life-breath and their functions, now offer all my actions and their fruits to the fire of Brahman (God). No matter what I may have done, said or thought in waking, in dreaming or in dreamless sleep, with my mind, my tongue, my hands or other parts of the body – may all this be an offering to Brahman."

It is very difficult to completely destroy selfish instincts and impulses all at once. Forcible repression is not advisable, as it would lead to injurious results. Therefore, by gradual sublimation, conquer the instincts. How is this to be achieved? Make it a regular practice to offer mentally all your actions and their fruits to God as in *Homa*. Such a practice will purify your heart. Whenever passions arise you should direct them to God, and thus remove the stings from the passions. For example, whenever you get angry, direct it against impediments to love of God. Ultimately you succeed in surrendering yourself totally to God.

GIVING UP THE FRUITS OF ACTIONS

Total surrender to God and inward renunciation of all worldly and religious concerns does not mean abandonment of actions beneficial to the world. As an instrument in Divine hands the devotee continues to perform actions, giving up their fruits to the Lord. (*N.B.S.*, 62). Lord Krishna, in the *Bhagavad Gita*, says:

"Whatever thou doest, whatever thou eatest,
Whatever thou offerest in sacrifice, whatever thou givest,
Whatever thou practiceth as austerity, O Arjuna,
Do it as an offering unto Me." (9. 27)
"Thus shalt thou be freed from the bonds of actions
Yielding good and bad fruits; with the mind steadfast in
The Yoga of renunciation, and liberated, thou shalt come unto Me. (9. 28)

Consecrate all actions to the Lord. Perform actions without any sense of doer-ship and without any attachment to actions and their fruits. Then you are freed from the bondage of Karma. You attain freedom in action. Your actions thus become selfless service and worship of the Lord. You live for the Lord only. You work for the Lord only. When actions are dedicated to the Lord, there is no rebirth for you.

In this path of self-surrender, all actions, all results and all rewards go to the Lord. There is no separate life for you. Just as the river joins the ocean and abandons its name and form, so also your soul joins the Supreme soul, giving up your name and form. You attain liberation, while living, and when your body falls you become one with the Lord. This shows the dynamic character of Divine Love as conceived by Narada.

Take, for instance, the lives of great teachers like Krishna, Buddha, Jesus Christ, Ramakrishna and others. (For details, *see* chapter 33. World teachers and Great devotees). They had realized God; they became One with God. There was nothing more to be achieved by them. Yet they continued to work for the benefit of the society without any personal or selfish motive.

REFLECTING UPON THE SCRIPTURES

A devotee must read devotional literature, reflect upon its teachings and follow them so that Devotion to God may become intensified in his heart. (*N.B.S.*, 76).

In Hinduism the *Vedas* and the *Upanishads* are ancient scriptures. The latter embody eternal truths relevant to all ages and all people. There are many other devotional scriptures devoted to the Path of Devotion. They include *Bhagavad Gita, Ramayana, Mahabharatam, Bhagavatam* and other *Puranas.*

The *Ramayana* deals with the glorification of God in the form of Rama, while *Bhagavatam* glorifies Lord Krishna, an Incarnation of God. *Mahabharatam* narrates the stories of kings, warriors, Sages and Saints, and the role of Krishna in establishing righteousness. The *Bhagavad Gita*, which is a part of *Mahabharatam,* is devoted to the teachings of Yogas. It is supremely popular not only in India but also among people devoted to Yoga all over the world.

Other devotional scriptures include *Vishnu Purana, Adhyatma Ramayana, Narada Bhakti Sutras, Sandilya Bhakti Sutras, Siva Purana, Devi Bhagavatam, Narada Pancharatra, Bhakti Rasamrita, Ananda Lahari, Periyapuranam, Tiruvachakam,* and the songs and writings of Mira, Kabir, Surdas, Tukaram, Alvars, Nayanmars, Sri Chaitanya, Saint Ramalingam, Sri Ramakrisha Paramahamsa, Ramana Maharishi, and all the Saints and Sages of more recent times.

In Buddhism, Christianity, Islam and other major religions there are basic scriptures, and teachings of Sages and Mystical Saints. The central theme of all scriptures and devotional texts is the enfoldment of devotion in the human heart. They inspire the devotees and sustain their Love of God.

A devotee, therefore, must choose the scriptures most inspiring to him, study them without flaw, reflect upon their teachings and follow them so that Devotion to God may be intensified in his heart. The *Thirukkural,* an ancient ethical scripture in Tamil, beautifully describes the process of such learning:

"Learn well what should be learnt, and then
Live your learning." (391)

This means that you should learn only that, which should be learnt, not anything and everything in writing. Life is short; so you should not waste your time in reading things, that spoil your mind. Learning is not a mere pastime. You should study worthy writings without flaw. Contemplate on what you have learnt. Learn good things and live according to them.

The study of books by the devotee involves three important implications. (new paragraph)

Firstly, a devotee has to be very selective in reading. He has to take care to read only such books as will nourish his intense Love for God. He does not need to dabble in such religious books as publications of the so-called rationalists and modern psychologists.

Secondly, even in the study of genuine devotional books, the devotee does not swallow everything that the books say. Like the bees interested only in honey, he has to take only the essence and reject everything else. He has to be awake to poetic exaggerations. Thirdly, the devotee has to be careful in the adoption of spiritual practices referred to in the books. He must follow only such practices as are conducive to devotion with reference to his background and circumstances. He should not blindly copy other great sages. He should adapt only such practices that he is capable of understanding and following.

The *Bhagavatam* (12. 13. 8) and *Narada Bhakti Sutras* insist upon the necessity of reflection on the teachings of scriptures. Through such reflection the devotee assimilates the teachings and follows their essence in his life. The teachings of great scriptures have inspired many people and they have become great devotees. Hence reflecting on the teachings of scriptures is very important.

27

Divine Grace

Introduction

"Success in Devotion (or spiritual life)" says Sutra 38 of *Narada Bhakti Sutras*, "is attained chiefly by the Grace of Enlightened Ones, as well as by Divine Grace."

When the spark of aspiration is kindled in the heart of an aspirant, it is said that he has received the Grace of God (*Ishvara Kripa*), he has been blessed by God to follow the spiritual path that will lead him to Self-realization or God-realization.

When an aspirant of Yoga is able to withstand temptations and overcome obstacles in a miraculous manner, it is said that he has perceived the Grace of his Guru, Spiritual teacher; he has become sensitive to his Guru's spiritual influence.

God's Grace is the sole Reality and Guru's Grace (*Guru Kripa*) is an aspect of Divine Grace, because God's manifestation is greater in a realized Sage than in anybody else.

While the previous Sutras highlighted the need for self-effort (spiritual discipline), Sutra 38 asserts that without Grace it is impossible to advance on the spiritual path. What is the relative importance of Grace and self-effort in spiritual life? There are different viewpoints on the relative roles of Grace and self-effort. Some argue in favor of Grace, some in favor of self-effort, and some in favor of a blend of both Grace and self-effort.

In Favor of Grace

We know from actual experience that many of our self-efforts do not produce even worldly effects desired by us, unaided by outside help. How can we then expect that spiritual Realization, which is so rare, can be attained by personal effort alone? According to some devotees, even the aspiration to strive for Self-realization is due to God's Grace.

This view is advanced to correct the aspirants' common error based on a misunderstanding of self-effort. Many aspirants cling to their egoism, pride and conceit in the name of self-effort. Therefore, in order to infuse humility and to show the need for self-effacement, Sages have highlighted Grace as the ruling force in human life. An aspirant, however, must purify his mind and heart by prayer, meditation, surrender and other spiritual disciplines and make himself fit to receive God's Grace. The egocentric attitude of a person can never attract God's Grace. In the words of Sri Ramakrishna,

> "Divine Grace, the healing and illuminating energy that rains down ceaselessly upon the human mind, heart and soul, cannot be absorbed by the high, rocky hill of personal interest and personal importance. This precious life-giving water runs off the high ground of ego, without ever penetrating its hard, barren soil." (*Great Swan: Meeting with Ramakrishna*, p. 30)

The subtle forms of lust, anger, egoism, pride, etc. can only be totally destroyed through the Grace of the Lord. However hard you may strive and do spiritual practice, these subtle forms cannot be eradicated without the Lord's Grace. However, you should not sit idle and say, "Lord's Grace will do everything for me. Why should I do any practices?" This is a wrong philosophy. God helps those who help themselves. God's Grace will descend on only those persons who take refuge in God.

Mira abandoned everything. She renounced kingdom, husband, relatives, friends and property. She remembered her Lord Krishna, all day and all night. She shed tears of devotion. She sang His praise with single-minded devotion. She gave up food. Her body got emaciated. Her mind was ever absorbed in Lord Krishna. Only then, did Lord Krishna shower His Grace upon her.

The *Mahabharatam* says that the revelation of God to man is the greatest of His boons to him, and that only he can see God to whom He is gracious, and not he who relies on self-effort alone. (12. 337. 20 and 340. 16. 17). This does not mean that the Lord is partial to human beings. Just as the lamp in a hall sheds light on everyone in it without making any distinctions, so also God showers His Grace equally on all; but through their own fault, many do not take advantage of it. It is not the fault of the sun if one does not take advantage of it by being shut up in a room. It is one's own Karma that enables one to secure Divine Grace. By making efforts in improper direction one is deprived of Divine Grace; but when the obstacles on the path of Grace are removed by fresh efforts in the right direction, he gets Grace once again. Thus self-effort is still needed to remove the hindrances brought about by past deeds; but primarily it is the Lord's Grace that saves.

In Favor of Self-effort

Just as the extreme school of devotees say that even self-effort is made possible only by the Lord's Grace, the extreme school of Jnanis, on the other hand, stress that Grace is only a pious imagination of the emotional devotees. According to them, it is only the weak and mean-spirited that rely upon external help, throwing the burden of their own responsibilities on the shoulders of someone else.

One can make or mar himself without any external help, for he alone is the architect of his fortune. They say, "no God can help anybody against himself; for, if it were otherwise, consistent with His gracious nature, God would have saved every sinner long ago".

The Vedantic scriptures such as *Yoga Vasistha* speaks of the supreme importance of self-effort. There are two reasons for emphasizing the importance of self-effort. *First,* most people renounce their self-effort thinking that everything depends upon God's Grace. In order to correct this fatalistic notion *Yoga Vasistha* has laid great emphasis on self-effort.

Secondly, the Self in the aspirant is in reality the Absolute Self (Brahman), and therefore, from a broad point of view, the entire process of spiritual evolution from bondage to Liberation is to be viewed as a stream of self-effort. The little self flows like a river to enter the Ocean of the Great Self.

A Blend of Self-effort and Grace

As long as an aspirant is steeped in ignorance, he needs to be awakened from the state of laziness and inertia by being inspired to exercise his self-effort without becoming egoistic, and seek Divine Grace. But as the aspirant advances, he begins to discover that the invisible Hands of the Higher Self in him are leading him to advance on the spiritual path. Therefore self-effort leads to Grace, and Grace intensifies his self-effort. This process continues until his personal effort merges into the Ocean of Divine Grace.

The different views on the relative roles of Grace and self-effort are not contradictory to each other. They are relevant to aspirants in different states of spiritual evolution.

Saint Nammalvar is reported to have asked God why all men are not saved by His Grace, and to have got the reply from the Lord that since man is endowed with a free will, God would wait to see man's predilection for Him. "Ask and it shall be given to you, seek and you shall find, knock and it shall be opened unto you." (Mathew, 7.7). Thus, although at first sight there is an irreconcilable opposition between the doctrine of self-effort or free-will and the doctrine of Grace, writers of Bhakti everywhere have found no difficulty in reconciling both to their satisfaction. They treat Grace as only an exception to the universality of the Law of Karma. All accept both the doctrines and allow each its own sphere of influence.

Spiritual realization is primarily due to the Grace of God working through Saints, but made available to us by our own effort. Even though we may get a natural inclination towards a higher life as a result of our spiritual practice in a previous life, proper stimulus from outside is almost necessary to stir up these latent tendencies. To many, this first awakening comes only through another perfect soul, a Guru. Even the Incarnations, Prophets and Messiahs had their first awakening from contact with such Gurus.

An aspirant should, therefore, be on the look out for a perfected soul and when he finds him, accept him as his Guru and seek his guidance. But it is quite possible for him to come under the spiritual influence of another without any deliberate effort. The perfected soul is one who attained union with Brahman. He is a man of God. The *Upanishad*s say: "Verily a knower of Brahman has become one with Brahman. To obtain the grace of such a Guru is the same as having the Grace of God." In the words of Sri Ramakrishna,

> "There is but one Guru, and he is Satchidananda, that is God, who is the immortal Being, the pure Consciousness, and abiding Love and Bliss. Human Gurus are like so many pipes through which the water of the same lake pours."

The Guru is one who opens the Divine sight of the disciple. Through the grace of the Guru, which is the same as the Grace of God, an aspirant attains Supreme Love and union with Brahman. (For details, *see* chapter 12. Guru's Role). God's Grace comes in the form of self-effort, spiritual awakening, yearning and striving. We then feel a tremendous urge to exercise our will-power or determi-nation to blast away all the obstacles in the path.

In the initial stages, mighty effort of sustained spiritual discipline is necessary for conquering the monkey mind and senses. This is why Lord Krishna in the *Bhagavad Gita* says,

> "Indeed this restless mind is difficult to control,
> but by practice and dispassion it can be controlled." (6. 35)
> "The Yogi who strives with great effort to purify the mind
> reaches the highest goal." (6. 45)

Thus, self-effort is needed to purify our minds and put our will in turn with the Divine Will. When we accomplish this, then we find that everything happens by the Divine Will. Then the apparent conflict between self-effort and Divine Will cease.

God's Grace manifests itself in various other ways also. Grace always gives the devotee a wonderful inner poise and strength to face all trials and difficulties and makes him purer and enables him to feel the Divine Presence.

God has revealed supreme knowledge through the scriptures like the *Vedas, Upanishads, Bhagavad Gita, Bible, Koran*, etc. to enable us to know and adopt the proper spiritual practices for realizing Him. He Himself comes down and takes human form for teaching righteousness (*Dharma*) and for providing us with a realizable ideal. The provision made in the Divine scheme for cancellation of all previous Karmas as a result of continuous meditation or devotion is another aspect of His Grace. The very inexorable Law of Karma is only due to His Grace; for He thereby gives ample freedom to individuals to work out their own salvation. In fact all the troubles and tribulations of worldly life are meant only to wean those who go astray, from their evil ways. Everything that happens is due to His infinite Grace and is meant only to bring us nearer and nearer to Him.

Above all, the most important manifestation of God's Grace is the inner urge for perfection that is present in every one of us. This urge makes us discontented with everything worldly and thus leads us higher and higher, step by step, until we reach the goal in the long run. That it is the Lord's Grace that saves us is clear from the life of many great devotees.

Scriptures are abound with stories of miraculous transformations in human nature brought about by the Grace of Sages. Even many wicked persons who approached Sages with the intention of hurting them became transformed into Sages themselves. Sages and Saints are like sandalwood trees that give their aroma even to the cruel blades of the axes meant to destroy them.

Sage Valmiki, for example, was originally a daylight robber. But by the grace of a Sage whom he planned to rob, he was drawn to the Path of Devotion and became transformed into one of the greatest Sages by his intense spiritual practice. He became the author of the great scriptures *Ramayana* and *Yoga Vasistha*.

28
Practice of Bhakti in Action

Prayer and meditation at fixed hours are most essential for spiritual progress, but that is not enough. The prayerful and meditative mood must be maintained throughout the day. There must be an undercurrent of thought about God even when we are engaged in our daily duties of life. It would prevent the rising of impure thoughts, and help us maintain the purity of mind.

Keeping the Mind on the Divine

It is important to keep the mind on God. The mind should never be allowed to drift away. But when we are absorbed in worldly affairs, we forget the Lord. Work and duties are unavoidable. It is possible to keep the mind on the Divine while we are working? This is possible only when work itself becomes a form of worship. Unless this happens, a little *Japa* or prayer and meditation done in the morning and evening does not help much. The only way of converting all daily activities into worship, in order to keep remembering the Lord always is to connect everything we do and think of what the Divine.

Sankaracharya says in a song: "Whatever I do, O Lord, is Thy Worship." (*Sivamanasapuja*, 4). Develop the attitude of doing everything as an instrument in the hands of God, and consider everything as an instrument in the hands of God, and consider everything as Divine work. Anything done in that spirit of consecration becomes a form of worship, and it is as (removed 1 word) effective as prayer and meditation.

Repeat the Lord's Name and think of Him during your idle moments. To the extent you succeed in this practice, a tremendous spiritual transformation will take place in you and you will be feeling the Divine Presence, Love and Bliss in your heart more and more. It is very dangerous to let the mind remain idle or to brood over the past. If you allow the mind to remain unoccupied, it will go on indulging in useless thoughts and brooding over past experiences. All these will only pull you down.

If you get into such a mood, at once take up reading some good book or do some selfless service to somebody. Then that mood will soon pass. Otherwise, if you just sit and daydreaming or brood over the past, you would not only waste time but also produce self-created obstacles. The importance of study of scriptures in spiritual life needs no emphasis. It is very important for most of the aspirants. Study and selfless work are forms of spiritual practice.

Sustained Effort

The quality of our spiritual life is determined by the type and frequency of the thought of the Lord that arises in our mind in the course of the day. In the midst of our daily work and duties, the thought of God must constantly bubble up. That is real spiritual life. But of course this calls for conscious struggle over a longer period of time. Unpleasant situations cannot be avoided in mundane life, and they disturb our mind. However, situations themselves are not the real cause of our mental disturbances. It is our reactions to situations that are responsible for our disturbances, and our reactions are the offshoot of our ego. For example if someone uses harsh words against us we immediately feel offended and retaliate, because of our egoistic sense of pride. Egoism is the major problem in spiritual life. Through prayer, *Japa* and self-surrendering to God we have to make concerted efforts to overcome the sense of egoism. (For details, *see* chapter 25. Cultivation of Bhakti: Positive Practice – IV.) Simultaneously we have to overcome our tendency to react adversely to situations in our family and social life by consciously developing a balance of mind. We should neither feel elated when someone praises us nor feel depressed when someone blames us.

Once we develop equanimity we could remain unperturbed by such pairs of opposites as pleasure and pain, profit and loss, and praise and blame.

Continuity of Practice

Further, ups and downs in spiritual life are inevitable. We should not worry about our mood, or mental ups and downs. As we go on steadily practicing *Japa* and meditation and feel the presence of the One that never changes, we will have greater stability. The Infinite Spirit dwells in our heart and in other beings as the Supreme Spirit. If we were able to feel more and more of the Divine Presence within us, we would gain greater mental stability.

Inner Contact with the Divine

Every devotee must have a definite center of consciousness, a holy Form (Chosen Deity) and a holy Name for constant use. Just like the mariner's compass constantly points toward north, our mind must be fixed always on the holy Form and the holy Name in our center of consciousness. This can be achieved over a period of time by regular practice of meditation in the mornings and evenings.

We have to visualize the Form of our Chosen Ideal (Personal God) at our center of consciousness. When we do not feel the living presence of our Personal God, our own consciousness loses its base and we will feel as if we were floating in the air without any support. Sometimes we succeed in making the form of Ideal very vivid during our mental worship or meditation, but we may not be able to establish a proper contact with it. This will bring much emotional disturbance. We should not lose heart, but go on praying with intense love. Then God's Grace will restore the inner contact. Actually our souls are always in contact with the Divine, but since we are not in the higher plane of consciousness, we do not feel it. Our impurities, impulses and fantasies keep us from rising to that plane of consciousness where we can easily feel the Divine contact. Hence the importance attached to the purification of mind in spiritual life. If we cultivate intense love for the Lord, our minds get purified. Then we tend to feel His presence in everything.

Over-activity

We often tend to go on multiplying work until it absorbs all our attention and energy. We should not run after work. We need a certain amount of leisure. We should always find time for our spiritual practices and devotion.

Over-activity is an obstacle to spiritual practice. It is an unbalanced mentality. Over-active people tend to lose balance and indulge in some sort of half-mad random activity like the restlessness of an intoxicated monkey. Then they complain, "Where is the time to do my spiritual practice?" Such restlessness will undermine both physical health and spiritual effort.

We should discriminate between real duty and unnecessary work like unimportant social functions. When we eliminate unnecessary work and streamline our regular duties, then we can find more time for prayer, *Japa*, meditation and scriptural studies.

Work as Worship

Ordinarily we find some difference between work and worship. This difference disappears, and work becomes worship, when we work as an instrument in the hands of the Lord. We should also repeat a *mantra* mentally, even when engaged in routine work. Then our whole attitude changes; work done in a spirit of dedication or as an instrument of the Lord becomes worship. Our whole life then becomes an undivided offering to the Lord. Lord Krishna, in the *Bhagavad Gita*, says that all beings have evolved from the Supreme Spirit that pervades everything, and that one can attain Enlightenment by worshipping It with one's duty (18. 46). Any honest work, however humble, may be looked upon as God's work. Really speaking we are not the doers; God is the only doer. His Will, His energy is working through our bodies and minds and through everything in this universe. When our will becomes one with the Divine Will, we will never become mechanical. On the contrary our very life becomes a Cosmo-centric life instead of an egocentric one. Whenever any opportunity for service arises, we should take it up with heart and soul. We should render service dispassionately in a spirit of dedication to the Lord. Selfless service to living beings is service to God, as God resides in all beings.

Intense Spiritual Effort

Through intense spiritual effort we can create an inner current in the mind that flows toward the Divine even when the rest of the mind is engaged in work. Through a steady spiritual practice we will be able to have a larger part of our mind under control. We can then feel more effectively the Presence of God in the midst of our duties.

Attitude toward the World

Our attitude towards others depends on our attitude towards ourselves. How do you look upon yourself? If you look upon yourself as a body, you see only human bodies all around you. If you look upon yourself as the luminous Self inhabiting the body, you will see the same Light shining in all beings around you. Then all external differences in the name of color, sex, caste, creed or religion will disappear, and you find in others the same Self within your body. Then your outlook would be totally different than that of an ordinary person who identifies himself with his body. You will see the unity in the midst of diversity of names and forms.

There are some devotees who look upon all men as God and all women as Goddess, and then finally transcend them, reaching the Supreme Spirit out of which they have come. Thus such devotees solve many problems. What a difference it makes when one looks upon all women as Goddess and all men as God!

If we are really able to look upon these physical forms – men and women – as a manifestation of the Formless and never as primary realities, if we are able to look upon matter as a manifestation of thought, and thought as a manifestation of Cosmic Consciousness, then we can see everything in its right place, in its true setting. Then we are no longer deceived by the physical or mental forms as such. Such an attitude is essential for spiritual life.

Similarly we have to solve the problem of good and evil. There must come a time when we see the Divine at the back of both good and evil. Then we rise above good and evil, and see the same Divine in everything.

GUIDELINES FOR PRACTICE

1. Consciously cultivate a very personal relationship with that form or aspect of God for which you have the most devotion. Stick to one form and one type of relationship: God as mother, father, child, friend, master or beloved.

2. Try to remember the Lord at all times. Let your first thought of the day, as you awake from sleep be that of God, and your last thought at night before you retire be that of God.

3. You can have pictures or statues of your beloved Lord to remind you of His presence.

4. Read and listen to stories of God's glory and play.

5. Chant God's Name and songs about His greatness.

6. Do Japa. As you repeat God's Name, think of your beloved Lord. Try to keep up the repetition throughout the day. Whenever your full attention is not needed for the task on hand, part of the mind can be repeating the mantra.

7. Do Puja. You can worship the Lord on your altar with incense, lights, fruits and flowers. Or you can do mental Puja with concentration and deep feeling.

8. When you have the opportunity, visit holy places. You will draw strength and light from the beautiful, pure vibrations that have been built up there.

9. Think about the great benefits and gifts you have been constantly receiving. The more you think of them, the more you will feel grateful, the more you will feel devoted, the more you will feel love. Every day you can write down five blessings in your life. At the end of the week, review the list.

10. Never forget the good thing that you receive, however small it may be. If somebody does some harm to you, forget it immediately.

11. Cultivate the attitude: "I believe in God. I have given myself into His hands. Whatever takes place happens because He allows it to happen. And it is for my good. If it were not for my good, why would God allow it to happen?" This is true devotion and true faith in God.

12. When you are trying to cultivate faith, hundreds of things will come and disturb it. There will be tests. The more you trust God, the more you get tested. Trust and tests go together. Have sincere and heartfelt prayer. Keep praying, "God, please help me. Let me see the Truth always."

13. Never get discouraged or give way to despair. When difficulties come, trust in God's mercy and grace, and do not get disheartened by your own limitations. According to the *Bhagavad Gita*, God's grace makes the dumb eloquent and the cripple cross mountains.

14. Think of the great teachings from the sages and saints, and apply them in your life. Draw inspiration from the lives of saints and sages.

15. Be regular in your spiritual practices. Regular practice will purify your mind and help you to grow in faith and devotion.

16. Associate with fellow seekers for support and inspiration. Seek the company of wise men.

17. God is love. Love one and all. Do not hurt anything even by your thought. Do unto others, as you would want them to do unto you.

29

Dangers and Degeneration of Bhakti

The noble pursuit of Bhakti is not totally free from dangers. Devotees have to guard against some dangers. They are described below.

Sensuality

Bhakti, in its lower forms, may degenerate into material attachment and sensuality if the Highest Ideal is not meditated upon regularly and carefully. A devotee may become a slave to sensual desires. He must guard against the attractions of mundane life. Through the force of dispassion and meditation, the desires get suppressed. But through the devotee's practices, his senses are heightened, and he is liable to become sensual; hidden desires reemerge with greater vigor. Then mental disturbances arise. The *Bhagavad Gita* warns against this danger:

"The excited senses of even a wise man, though he be striving,
Impetuously carry away his mind. The mind that yields to
The roving senses carry away his discrimination,
Just as the gale carries away a boat on the waters. (2. 60, 67)

So the devotee should be ever vigilant and watch the mind. When any old desires come up to the surface of the mind, he must nib them in the bud by intensifying dispassion and discrimination. He must go into solitude, observe fasting and silence, and intensify the practice of *Japa* and meditation.

Bhakti may be heightened by such means as music and rhythmic dancing. However, most of those, who reach the state of ecstasy by such means, cannot control themselves when the ecstasy is over. They fall down from the height of their ecstasy as fast as they go up, like little rockets shot into the air.

Bhakti is an energy stored in our bodies and minds. It is eager to express itself. The energy within has to be controlled and directed to constructive channels of expression. Any misapplication of that energy results in degeneration.

The remedy for sensuality is creative art, and humanitarian service. Artistic expression is a great channel for the outflow of the emotions. Social work is suitable as an outlet for some people but it should be done in a spirit of service. Service to living beings is service to God.

Mental Reaction

We must beware of mental reaction. When we expect too much too soon, we are bound to run into disappointment. Our devotion to God is to be, just for the sake of love, and not for any expectations. Where there is no 'appointment,' there is no disappointment.
It is easy to gain ground, but difficult to hold it. One must become intoxicated with love for God, but the process must be well regulated. One must follow all the required steps carefully. When an elephant enters a small pool of water, what havoc is created! So the cultivation of right attitude and the expansion of consciousness should match with the intensity of devotion.

Fanaticism

The great disadvantage of the Path of Devotion is that in its lower forms it can often degenerate into hideous fanaticism. The singleness of attachment to one's own Ideal of religion, without which no genuine Bhakti can grow, is very often also the cause of the denunciation of everything else. The undeveloped mind in every religion has only one way of loving its own Ideal; that is to hate every other Ideal. The fanatic loses all power of judgement. The same person who is kind and loving to people of his own faith does not hesitate to hate and even to kill people belonging to other religions. One may worship Krishna, another may worship Christ, and the third may worship Allah. Each one worships the Ideal that suits his bent of mind and background. The fanatics do not realize this truth. Themselves blind, they want to keep others also blind. They fail to realize that all the great Incarnations and Prophets of world religions are expressions of the same God. There is no one Savior for all. To force one's Savior on all is a most dangerous idea.

Nevertheless, the danger of fanaticism exists only in the lower stage of Bhakti. Once Bhakti becomes ripe and develops into a Supreme Devotion, there is no more fear of hideous manifestations of fanaticism. The matured devotee realizes that the different Ideals are the expressions of the same God. He has a universal vision, because he sees with the eyes of God. A wise person will easily see that different approaches to the Truth are necessary to suit the different tastes, temperaments and customs. This is wisdom.

Fanaticism is not restricted to worshippers of Incarnations and Prophets alone. Even non-dualists can become fanatics. To proclaim to everyone that gods, goddesses, and Incarnations are all illusory, being products of Maya, is the height of foolishness. Even the non-dualists must realize the Divine manifestation in all creation. For many people the manifested God is as important as the un-manifested Absolute. The way to the transcendental Absolute lies through the Immanent. First we must realize the Divine within ourselves, then in others, and finally transcending all names and forms, we should realize the One without a second.

As the seeker advances in spiritual life, he finds that the Form that rises out of the Formless merges back into It. Love for the Personal God then changes into love for the Impersonal Absolute. The true devotee sees everything in proper light. He knows that the form and the Formless, the Personal and the Impersonal are but two aspects of the same ultimate Reality. Having risen above the clutches of worldly attachments, he loves God in all His various manifestations. There are no boundaries to his love.

False Contentment

The devotee gets some experiences in the course of his practice. He may see wonderful visions. He may hear various melodious sounds and get some powers as thought reading, foretelling, etc. He then foolishly imagines that he has reached the highest goal, and stops his further practice. He gets false satisfaction. This is a serious mistake. Such experiences are auspicious signs that manifest on account of a little purity and concentration. They should just serve as a source of conviction and motivate the devotee to intensify his practice.

Spiritual Pride

As soon as a devotee gets some spiritual experience or psychic power, he may be tempted to be puffed up with pride. He may boast and brag too much of himself. He separates himself from others and treats others with contempt. He may say: "I have practiced unbroken celibacy for the last twelve years? Who is pure like me? I lived on mere water for 30 days. Who can do this? I have done service in an ashram for ten years. No one can serve like me?" This kind of pride is a serious threat to spiritual progress. The devotee should guard against this danger and develop humility through total surrender to God.

Superstition

With reason and logic, religion can easily degenerate into the worst kind of superstition. There is nothing mysterious about religion or about Bhakti. You should not become a slave to dogmas and rituals. You have a right to ask "why" and "what for." One needs the stimulus of good company to prevent oneself from brooding and to keep the mind alert.

Sanctimonious people are often superstitious. They repeat scriptures and hymns parrot-like without following their spirit in their life. Religion has to be lived.

Temptation of Psychic Powers

Another danger to be guarded against is the temptation of some psychic powers that may develop in the later stages of Bhakti. After a devotee practices devotion for some years, he may attain some occult or spiritual powers. He may become enamored of such powers. They drift him away from the ultimate goal of spiritual practice. They are nothing but obstructions to his growth. With the help of the Guru he should overcome such temptations and become indifferent to psychic powers.

In spite of the possibility of the above dangers and obstructions on the Path of Bhakti, it is a very natural way to reach perfection. By one-pointed Devotion to the Lord, the devotee may realize the highest goal, provided he is sincere, humble, pure and selfless.

30

God- Realization

Introduction

Spiritual life is a continual war between our higher and lower natures. Animal and Divine tendencies are found, often mysteriously blended, within each one of us. Animal tendencies or demoniacal qualities intensify our ignorance and bondage. Divine virtues, on the other hand, lead us towards Perfection and Liberation. It must be the constant effort of a devotee to eliminate the evil and strengthen the good in himself. The *Bhagavad Gita* enumerates these Divine virtues and demoniacal qualities in detail thus:

> "Fearlessness, purity of heart, steadfastness in knowledge and spiritual practice, giving away in charity, control of the senses, sacrifice, reading of the scriptures, austerity and uprightness; non-injuriousness, truthfulness, absence of anger, renunciation, tranquility, absence of calumny, compassion to beings, non-covetousness, gentleness, modesty, absence of fickleness, boldness, forgiveness, fortitude, purity, absence of hatred, and absence of pride – these belong to one born in a Divine state. (16. 1-3)
> "Hypocrisy, arrogance and self-conceit, anger, harshness and ignorance belong to one who is born in a demonical state. (16. 4)
> "The Divine nature is deemed for Liberation, the demonical state for bondage." (16. 5)

Supreme Devotion and God-realization

The Knowledge of Self liberates the bound soul from the vicious cycle of births and deaths, and grants it unchanging and eternal bliss and peace. It is then Supreme Devotion (*Para Bhakti*) dawns on him. This is the summit of God-realization. The devotee then beholds the entire manifested Universe as the supreme image of his beloved Lord; he looks upon all creatures and things in the Universe as the very form of his Lord. He feels and sees the Lord's presence everywhere; the sense of duality completely vanished in him. The Oneness of all existence becomes now the keynote of his vision. In this supreme state the devotee enjoys unending and inexpressible bliss. There is no limit to his Divine ecstasies. He veritably rolls and swims in the infinite ocean of joy. Blessed is the devotee who reaches this highest peak of God-realization.

God-realization and World Religions

All great religions of the world recognize the possibility of this highest spiritual experience. This is called by various names by different philosophers and Sages of different countries. In Raja Yoga it is called *Samadhi* (Absorption). The Buddhists call it *Nirvana*, which means the cessation of misery, sorrow and all imperfections, and attainment of Perfection. The Christian Mystics of the Middle Ages described it as Ecstasy, and modern Christians call it the State of Communion with God.

Jesus became the Christ after attaining it, and Sakya Muni became the Buddha or the Enlightened after he attained Illumination. Sri Ramakrishna, the great Sage of the Nineteenth Century in India, reached that stage and is now worshipped by thousands of people as an Incarnation of God upon earth. Many philosophers have attained to this state of Divinity. Plotinus, the Neo-platonist who lived two centuries after Christ, reached it four times in his life. Porphyrious attained to this state when he was 66 years old. Dionysius, who lived in the Fifth Century, called it the State of Mystic Union, the Union of the soul with God. The great Christian mystic Meister Eckhart, who lived in the fourteenth century, described the nature of this state thus:

"There must be perfect stillness in the soul before God can whisper His word into it, before the Light of God can shine in the soul and transform the soul into God. When passions are stilled and the worldly desires silenced, then the word of God can be heard in the soul."

Stages of God-realization

There are two stages of spiritual experience. The devotee first experiences the vision of his Chosen Ideal or Personal God accompanied by inexpressible bliss. There is still a sense of separateness from God. But when he reaches the highest stage, Love, Lover and the Beloved all become one; there is complete union with God. Then God is experienced as impersonal, immanent and transcendent Consciousness or Supreme Self.

How does a devotee reaches this Supreme State? He accepts one particular form of God as his Chosen Ideal and begins to think and meditate on that alone. In the beginning he fails to bring before his mind's eye in meditation the complete form of his Chosen Ideal (*Ishta*). As he continues his practice of meditation, his mind gets more concentrated, and his mental image of God also becomes more vivid and steady. Subsequently, in a profound state of meditation, the image becomes living, and it becomes possible for the devotee even to feel its Divine touch and hold talks with it. At this stage, whether with eyes open or closed, the least concentration makes the devotee feel the living presence of his Lord. Then again, from the belief that it is his Lord who has assumed all the different forms, the devout devotee perceives the emanation of various other Divine Forms from his Chosen Ideal. Bhagavan Sri Ramakrishna used to say, "To him who is blessed with the vision of one such living Divine Form, the vision of other forms of God comes easily, by themselves."

Though these visions have their inception in the subjective mind of the devotee, yet in the matured state of meditation, when the vistas of the thought-world are opened up to the mental gaze, they assume all their true colorings of objective reality, as experienced in the waking state. As these Divine visions begin to come more frequently, and the consciousness of the unreality of thought-world grows deeper, scales fall from the devotee's eyes. He begins to perceive that the so-called objective world, with all its apparent concreteness, is but a fruit of the mind's imagination. Again, the visions in this state of deep meditation appear with such all-absorbing reality, that for the time being the objective world is completely obliterated from his consciousness. This state has been described in the Scriptures as *Savikalpa Samadhi*. In that state of consciousness, though the material world disappears altogether, there still remains for him the other world, the world of thought. Through it the devotee feels the joy and sorrows of life with his God, exactly as we feel in our relationship with the persons of the outside world. The only difference is that the web of all his emotions and desires is woven around his Beloved Lord. This is a state where mental activity is not completely stopped but is held fast to one idea only, namely God.

Thus, by meditating on one object of the thought-world, the perceptions of the gross, external world disappear altogether, and when one idea gets hold of the mind strongly, all other ideas fall off from it. The devotee who has advanced so far is not far from the state of Self-realization or what is called *Nirvikalpa Samadhi*. It is needless to point out that he who has succeeded in getting rid of the idea of the reality of the world has got his will sufficiently strengthened and his mind sharpened. As a result, he soon comes to learn that if he can bring about a complete cessation of all mental activities, his enjoyment of the Divine Bliss will be much more intense than hitherto, and so his mind eagerly runs toward the realization of that state. Through the Grace of God and his Guru, he too soon crosses the thought-world and realizes the Absolute Unity and thereby attains eternal bliss. Or, it might be that his intense love for God itself takes him ultimately to that state. Then he feels like the Gopis of Brindavan, his identity: One with the Beloved. (Source: Swami Vivekananda and others, *The Message of Our Master*, Calcutta: Advaita Ashrama, 1968). This state of Self-realization is called Perfection, Liberation, Freedom or Immortality.

Perfection

When Supreme Devotion for God arises in the heart of a devotee he attains Perfection. (*N.B.S.*, 4). Perfection means Oneness with

God. The individuality of the devotee merges within the Supreme Divinity. It is the goal of life. Strictly speaking, this Perfection is not something to be attained, for, the true Being – the Self or Atman – is one with Brahman already. Perfection is already inherent in Man. It is only clouded by ignorance. Ignorance obstructs our Divine sight. It is the sense of ego that arises through the identification of the Self with the non-self, namely, the mind, senses and body. When the ego is transcended, the indwelling God is realized as our very Self, just as the sun shines when the clouds clear away. The function of spiritual practice is only to remove this cloud of ignorance or *Maya*.

Actions performed without any personal motive become selfless work. Selfless work purifies the mind and makes it fit to receive the Light of God.

Liberation or Freedom or Salvation

Salvation is concerned with the real nature of the soul and its ultimate destiny. Every religion has its own concept of Salvation, but all of them agree that it is a state of perfect happiness that the soul attains ultimately. Hinduism holds that Salvation means *Mukti* or Freedom. According to Judaism, Salvation can be obtained by leading a perfectly moral life; to this, Christianity adds that one should have faith in Christ as the only Savior. According to Islam Salvation is entirely in God's fiat, and faith in the Prophet Mohammed is absolutely essential to receive it.

According to Jainism, Liberation means the complete dissociation of the soul from matter. This can be attained by stopping the influx of new matter into the soul and by the complete elimination of the Matter with which the soul has already become mingled. The Karmas, or the forces of passion and desires in the soul, attract to it particles of Matter. The Karmas lead to the bondage of the soul through Matter. By removing Karmas through right conduct, a soul can remove the bondage and regain its natural Perfection.

Buddhists call Liberation *Nirvana*, the extinction of passions. According to Hinduism, Jainism and Buddhism, Salvation can be attained even in this life while one is alive. One who has attained this state is called a *Jivanmukta* (*see* chapter 32. Jivanmukta.) According to Christian theology Salvation is possible only after death. This is also more or less the view of Islam. However, a number of great mystics in these religions have attained the mystical experience of oneness with God while alive.

What is Salvation or Freedom? In social life we hear of freedom from want, freedom from fear, freedom of speech and freedom of religion. But all these are limited freedoms. All modern democratic states guarantee these freedoms to their citizens. Real freedom is the freedom of the soul. The soul of human beings is bound by the fetters of misery and the cycle of births and deaths. True Freedom is freedom from this bondage by becoming One with the Divine. In the awakening of a higher state of consciousness, the soul realizes its true nature as the Absolute Self, the ultimate Reality of the Universe, and attains Freedom. Buddha said, "By attaining *Nirvana*, the illumined soul is established in holiness and attains freedom." Christ said, "And ye shall know the Truth and Truth shall make ye free." The ideal of Vedanta, the most important School of Indian Philosophy, is eternal Freedom, a state of eternal peace and bliss. Bhakti Yoga and other forms of Yoga are a means for attaining this total Freedom in this very life. Our aim is to become free while living, in this very life. What is the use of attaining freedom after death?

Immortality

Self-realization is also known as Immortality. Immortality means freedom from death or mortality. Modern science proves conclusively the impossibility of complete annihilation. The very existence of a thing implies continuity, though its existence may be in a different form, under different conditions. Attainment of Immortality through the Knowledge of the Self or the Indwelling God does not mean continuity of existence within time and space. It means transformation and expansion of consciousness, and the realization that one is not the body or the mind, but the Self that is one with Brahman, the ultimate Reality. Advancing further, it is realized that Brahman alone is the only Reality. The Self is eternal and is beyond time and space. Time and space and other conditions of human life belong only to the finite world, and not to the Infinity.

Immortality does not mean attaining heaven (*swarga*). The heaven which all popular religions offer as a reward for good actions, is not permanent, nor is it a place of unalloyed happiness. The Hindu scriptures always emphasize the *impermanence* of heaven and celestial pleasures. There is no eternal heaven. In *Bhagavatam* we read:

> The man of sacrifices goes to heaven, worshipping the gods through sacrifices here below. Like a god he enjoys there celestial pleasures that he himself acquired....He enjoys pleasures in heaven till the merits of his good deeds are exhausted. Then, on the expiry of his merits, he falls down against his will, being propelled by time. (11. 10. 23, 26)

We also find the same view in *Bhagavad Gita*:

> "(All) the worlds, including the world of Brahma, are subject to return again, O Arjuna. But he who reaches Me, O Son of Kunthi, has no birth. (8. 16)

Once the effects of meritorious actions are exhausted, one has to be born again on earth. The *Upanishads* confirm that everything that is an effect produced by an action must pass away and cannot be permanent. On the other hand Immortality, once achieved, can never be lost. It is a state of eternal bliss. Worldly pleasure and celestial joy pale into insignificance when compared with this Divine Bliss (*Ananda*). *Upanishads, Bhagavad Gita* and *Bhagavatam* (12. 12. 51) emphasize the blissful nature of God-realization.

> "With the mind unattached to external contacts the Yogi finds bliss in the Self;
> With the mind engaged in the meditation of Brahman, he attains to The endless bliss." (*Gita*, 5. 21)
> "When the Yogi feels that infinite Bliss that can be grasped by the pure intellect,

And that transcends the senses, and established wherein
He never moves from the Reality." (*Gita*, 6. 21)
"Supreme bliss verily comes to the Yogi whose mind is quite peaceful,
Whose passion is quieted, who has become Brahman,
And who is taintless." (*Gita*, 6. 27)

Immortality can be attained in this very embodied life. Both Narada and Sankaracharya accept the realization of Immortality in this very embodied life. The following passages of *Upanishads* support this view:

"Verily, while we are here we may know this (Self); . . . Those who
Know this become immortal." (*Brahadaranyaka Upanishad*, 4. 4. 14)
"By heart, by thought, by mind apprehended, they who know the Self
Become immortal." (*Katha Upanishad*, 2. 3. 9)
"If here one knows It (Self), then there is truth, and if here he knows It not,
There is great loss. Hence seeing It in all beings
Wise men become immortal." (*Kena Upanishad*, 2. 5)
"Those who through heart and mind know the Self as abiding in
The heart become immortal." (*Svetasvatara Upanishad*, 4. 20)
"Verily, he becomes Brahman who realizes Brahman. He overcomes evil
And transcends grief. Being free from all knots of the heart,
He attains immortality." (*Mundaka Upanishad*, 3. 2. 9)
The true devotee never craves for Liberation; he is quite satisfied with enjoying the Love of God for love's sake, and serving Him for the sake of service. Yet Liberation comes to him by the Grace of God.

Katha Upanishad declares,

"This Self cannot be attained by instruction, nor by intellectual power,
Nor even through much hearing. He is to be attained only by the one whom
The (Self) chooses. To such a one the Self reveals his own nature. (1. 2. 23)

In a sense, all life is from God, all prayer is made through the help of His Grace, but the heights of contemplation that are scaled by a few are attributed in a special degree to Divine Grace. If one becomes aware of God's presence in one's soul, it is due to God's own working in the soul. St. Paul says, "Work out your salvation with fear and trembling; for it is God who worketh in you both to will and to do of His good pleasure." (*Epistle to the Philippians*, 2. 12-13). Liberation come even if a person keeps idle and does not deserve such grace by his self-effort? No. Divine Grace never comes until the mind is purified by continued acts of self-sacrifice. God's Grace descends always on a person like a breeze, but if he wants to take advantage of it, he must do spiritual practices, as the boatman must unfurl the sails before he can catch the breeze. (*See* Chapter 27. Divine Grace.)

The Power of God-realization

The Realization of God and the constant enjoyment of Divine Bliss make a person spiritually and morally pure and healthy, and fit to undertake any kind of hard work in the service of God and humanity. It makes the dumb eloquent, and the lame cross mountains, as a poet says; "The fool becomes a poet, and the weak and cowardly become heroes under its influence." Witness how Jesus, a carpenter's son, became the wisest of his age, and brave enough to defy the mighty Roman Empire; or how Prahlada dared to disobey his father whom all the world dreaded and obeyed.

Communion–even after God-realization

As Swami Ramdas describes in his *Divine Life* (1991), even after the Realization of God as the immanent and transcendent Spirit, God can still be to you an intimate, ever-present companion, friend and protector. This personal relationship with Him sweetens your life in a marvelous manner. His presence enthuses and guides you at all times. He makes you the vehicle of His infinite love and mercy. He uses you as His instrument for spreading peace and goodwill on earth. You are one with Him and at the same time you are His free and cheerful servant and pure and radiant child.

You may reach the height of His impersonal nature. You may dissolve your little self in His all-pervading and infinite Consciousness. You may behold Him everywhere, but communion with Him as a personal Truth and Ideal is a rare and exalted experience. Now you can converse with Him, play with Him and be ever joyful in His company.

In this state, you never feel lonely. He is your never-failing friend. He is not a person in the sense in which you see and feel about the forms of beings in this world. Both the devotee and God belong to a realm other than the one that is gross and material, p. 38.

31

Worldly Persons and God-Realization

Worldliness

Worldly people (householders) are entangled in worldliness. They have various duties and responsibilities. They have to work hard to earn the means of living. They have to take care of their family and extend support to distressed and needy people. Though they may be devoted to God, they do not give God even a very small part of the love they feel for their kith and kin and worldly objects. Even when they offer prayers or contemplate God or chant His name their attention is diverted to worldly problems.

Bondage to Worldliness

Worldly people are bound beings. They are attached to family and worldly objects. Desires and attachments to worldly objects are their primary obstacles to spiritual practice and liberation. Lust and greed bind these people and rob them of their freedom. Why are people bound like this? It is because of egoism—the sense of "I, me and mine." This egoism has covered everything like a dark veil. It makes a person attached to name, fame, position and property. Maya is nothing but egoism.

One can get rid of the ego after the attainment of Knowledge, but one cannot obtain Knowledge of Self unless one gets rid of the feeling that one is the body. In this age, the human life is centered on food. So one cannot get rid of the feeling that one is the body, and the feeling of the ego. Therefore the Path of devotion is prescribed for this age. One should pray to God with firm conviction and a longing heart.

God of Worldly People

But As Sri Ramakrishna says,

"Do you know what the God of worldly people is like? It is like children talking to one another while at play, 'I swear by God.' They have learnt this statement from the quarrels of their aunts or grand mothers. Or it is like God to a dandy. The dandy, all spick and span, his lips red from chewing betel-leaf, walks in the garden, cane in hand, and plucking a flower, exclaims to his friend, 'Ah! What a beautiful flower God has made!' But this feeling of a worldly person is momentary. It lasts as long as a drop of water on a red-hot frying pan."[1]

Worldly people's minds are not steady. Sometimes their minds go up. Sometimes they go down.

God-realization

Is it ever possible for householders to realize God? Is there any hope for persons bound to the world? Sri Ramakrishna says, "Certainly there is." From time to time they should go into solitude to meditate on God. Further, they should practice discrimination between real and unreal and pray to God, "Give me faith and devotion." Once a person has faith he/she has achieved everything. There is nothing greater than faith. Rama who was God Himself had to build a bridge to cross the sea to go to Sri Lanka. But Hanuman, trusting in Rama's name, crossed the sea in one jump and reached the other side. There is a story to illustrate the efficacy of faith.

Once there was a milkmaid who used to supply milk to an ashram. She came from a neighboring village and had to cross a big river to get there. So she would take a boat across the river and deliver the milk.

One day she was very late. The Swami in-charge was looking for some milk, and when he did not see any, he asked his assistant what happened. "The milkmaid hasn't come today," he answered.

"What's wrong?" asked the Swami.

"I don't know, we're still waiting for her," said the assistant.

Two hours later the milkmaid arrived, and the Swami casually asked her, "Why are you late today? You used to come so punctually."

"Oh, Swami, the river was in floods and the boatman couldn't cross it. So I had to wait till the river subsided a little, and the boatman felt confident enough to bring us across."

Then the Swami jokingly said, "Aha, what is this? You come here, bring me milk, and sometimes even sit and listen to the satsang. Just with my satsang, people cross the big ocean of life. Can't you cross this little river?"

"Oh, sorry, I never thought of it. I will remember it from now on." And she returned to her village.

The rain continued to pour, and the next day they expected her to be late, but she was punctual. And every day afterward she was punctual. And the Swami was wondering how she could to be punctual when the river was in high floods. So he questioned her, and she replied, "Swami, you gave me the trick. I know that now. I don't need to wait for the boat."

"What?"

"Yes, Swami, I don't need to wait for the boat."

"Then how do you come?"

"I just think of you, the Guru, and walk."

"You really walk on the water?"

"Yes."

"On the water!"

"No, the water comes up to the ankles, but it doesn't come any further. See, even my dress didn't get wet."

He became very suspicious. "Can I see you doing that?"

"Sure, why not, come on Swami. I have to go back anyway."

So they went to the riverbank. The milkmaid started walking across and called to him, "Come on, Swami, come on." He was little embarrassed and thought that something would happen, so he stepped into the water. And as he began to walk, the water started raising up around him. "Swami," she said, "with your name in mind I am walking, and you can't walk yourself?"

"Lady," he said, "I am only a ladder. I can make others go up, but I can't go. You have faith and are able to walk. But I don't have that pure, childlike faith." He walked back and felt so ashamed that he was just a scholar, that he had learned too much philosophy and had not realized the Truth.

We can not gain God by just talking philosophy. God is the simplest thing in life, and the nearest. He is even nearer than your own heart. A simple God should be approached in a simple way. The simplest thing is total Bhakti. Just have faith.

"One should weep for God. When the impurities of the mind are thus washed away," says Sri Ramakrishna, "one realizes God. The mind is like a needle covered with mud, and God is like a magnet. The needle cannot be united with the magnet unless it is free from mud. Tears wash away the mud, which is nothing but lust, anger, greed and other evil tendencies and the inclinations to worldly enjoyments as well. As soon as the mud is washed away, the magnet attracts the needle, that is to say, the person realizes God. Only the pure in heart see God."

The Mode of Living in the World

Sri Ramakrishna teaches us how to live in the world:

"Do all your duties, but keep your mind on God. Live with all — with spouse, children, father and mother — and serve them. Treat them as if they were dear to you, but know in your heart of hearts that they do not belong to you. Give God the power of attorney. Make over all your responsibilities to Him; let Him do, as He likes. Live with a detached mind.

"If you enter the world without first cultivating love for God, you will be entangled more and more. You will be overwhelmed with its danger, its grief, and its sorrows. And the more you think of worldly things, the more you will be attached to them.

"First rub your hands with oil and then break open the jack fruit; otherwise they will besmeared with its sticky milk. First secure the oil of Divine Love, and then set your hands to the duties of the world. The world is water and the mind, milk. If you pour milk into water, they become one, you cannot find the pure milk any more. But turn the milk into curd and churn it into butter; then, when the butter is placed in water it will float. So practice spiritual discipline in solitude and obtain the world, the two will not mix.

"God has put you in the world. Resign everything to Him. Surrender yourself at His feet. Then there will be no more confusion. You will realize that it is God who does everything. All depends on His will.

"It is not harmful for a householder to live with his wife. But after the birth of one or two children, husband and wife should live as brother and sister.

"Get rid of worldly attachment by practicing spiritual discipline. Perform your duties without attachment. Surrender the fruits of your work to God. Meditate on Him. The more you meditate on God, the less you will be attached to the trifling things of the world. The more you love the Lotus Feet of God, the less you will crave the things of the world. You will then get rid of your bestial desires and acquire godly qualities. You will be totally unattached to the world. Though you may still have to live in the world, you will live as a Jivanmukta." (Source: The Gospel of Sri Ramakrishna, 1986)

REFERENCE

1. Swami Nikhilananda, tr., The Gospel of Sri Ramakrishna, Madras: Sri Ramakrishna Math, 1986, p. 624

32

Jivanmukta (Liberated Sage)

Meaning

The devotee who attains the state of Liberation or Immortality, while still living with the physical body, is called a *Jivanmukta* (Liberated Sage). He achieves this Liberation or eternal Freedom when he gets rid of Ignorance and its paralyzing effects. The seed of Ignorance must be fully roasted in the fire of the Knowledge of the Absolute, so that it will never germinate again. Then alone, the devotee becomes a *Jivanmukta,* breaking through the fetters of attachment. Thereafter he remains in the state of Eternal Bliss or Super-consciousness at all times and in all situations. He lives in the body as long as the momentum of the past actions that have produced it endures. When his soul leaves the body, it does not reincarnate again but is absorbed in the Absolute Brahman. As milk poured into milk becomes one (with the) milk, so the illumined soul absorbed in Brahman becomes one with Brahman.

Jivanmukta rightly understands that the Self or Atman and the body are two separate things. After realizing God, one does not identify the Self with the body. These two are separate, like the kernel and the shell of a coconut whose milk has dried up. The Atman dwells, as it were, within the body. When the milk of worldly-mindedness has dried up, one gets Self-knowledge. Then one experiences one's real nature of Divinity and realizes that God alone is the Doer, and he himself a mere instrument in the hands of God.

The Characteristics of a *Jivanmukta*

What are the signs or characteristics of a *Jivanmukta*? How does he conduct himself? The *Jivanmukta* is beyond the ego, beyond the three modes (*Gunas*) of Nature, free from desires and selfishness. He is established in the Self. He sees the Self in all beings and all beings in the Self.

The characteristics of a *Jivanmukta* are described below.:

Purity: As a freed soul, a *Jivanmukta* is not fettered by evil tendencies and conventional moral codes. He transcends the scriptures and social conventions. He is beyond the imperatives of ethics. Yet he cannot do anything that is not good and not conducive to the welfare of others, because purity has become his essential nature. In the words of Sri Ramakrishna,

> "In whatever he does, his conduct is in full agreement with the highest ethics, and he is above the distinctions of relative good and relative evil."
> All ethical virtues that he practiced before attaining Liberation now adorn him.

Free from Egoism: The *Jivanmukta* is free from egoism and attachments. Egoism or the sense of "I, me, mine" makes one's mind complex and calculating. Its nature determines one's attitude toward others. With the attainment of the highest super-conscious state all the knots of the heart and angularities of the mind are cut asunder and all doubts are dispelled. *Mundaka Upanishad* declares,

> "The knot of the heart is cut, all doubts are dispelled and
> The wise person's deeds terminate when Brahman is seen." (2.2.9)

All moral conflicts disappear and the Divine harmony is seen everywhere. So the *Jivanmukta* is even-minded and friendly and compassionate to all. Having attained the Divine, he has become the Divine and as such one with the souls of all.

Unselfishness: Having fully attained the goal of life the *Jivanmukta* no longer lives for any purpose of his own. His body and mind being in tune with the Cosmic Power and Will, he lives only for fulfilling some Cosmic purpose. He may engage in active work for the good of others or remain in communion with God. Whether he works or not, he surcharges the psychical world with his intensive spiritual thoughts and fertilizes it for the benefit of others. His very presence is a blessing to mankind. He promotes the good of the world even through silence.

Unbound by Actions: A *Jivanmukta* is not bound by the effects of actions. The impressions of his past actions cannot produce any effect, because they were roasted in the fire of the Knowledge of the Absolute. He may continue to be active, but does not identify with his body and mind, and he has no sense of doership. His actions are the outflow of Divine energy in and through him. His work, as the *Bhagavad Gita* says, will be for the welfare of the world. It is not done with the awareness that he is bringing about peace and harmony in the world. The sun does not give light with a view to cheering us. We just take advantage of the light and are happy. So also the actions of a *Jivanmukta* are spontaneous. They are actuated by the Divine Power for the fulfillment of the Divine plan.

Equal Vision: A *Jivanmukta* realizes that all phenomenal manifestations are strung, as it were, on the one omnipresent, invisible and immortal God. In the supreme harmony, unity and equality of this vision, all sense of diversity and differentiation is dissolved. In the vision of equality, the *Jivanmukta* sees the same Reality in all beings. The distinctions of caste, creed, color and race have no significance for him. He looks upon the learned Brahmin and the so-called Untouchables with the same vision of equality. He beholds the entire universe as one dazzling image of God.

A *Jivanmukta* is beyond pleasure and pain, and other pairs of opposites, and is beyond all grief. (*N.B.S.*, 5). This does not mean that he has no sympathy for the miseries of others. How can a *Jivanmukta* who feels the whole world as himself be indifferent to the miseries of others? His heart is full of compassion. As Sankaracharya says, in his *Vivekacudamani*, "He works of his own accord to cure the miseries of others, just as the Moon of its own accord cools the earth when it is scorched by the hot rays of the Sun."

Free from Desires: A realized sage has no more selfish desires. (*N.B.S.*, 5). Desires arise from a sense of imperfection which is a characteristic of an egoistic person. The feeling of imperfection is possible only when a person finds something outside himself, an object of seeking. But the perfect sage is not aware of anything other than God, his own higher Self. So desire has no place in him.

It is the accepted creed of all spiritual disciplines that non-attachment to the objects of the world is an essential pre-requisite of all spiritual pursuits. Why then, is desirelessness is mentioned as a special mark of the *Jivanmukta*? There is a difference in the quality of dispassion of the spiritual aspirant and that of the perfected one who has attained the goal. As the *Jivanmuktiviveka* notes, "at the stage prior to the realization of the goal, the seeker of God is free from cravings, as a result of his vigilant practice of self-control and other virtues; the desires still persist and are held in control only with some effort, whereas, after realization, there being nothing like the transformation of the mind, desires cease altogether....He is pure within as the sky which remains pure." "One who has realized the Self is free from hatred, hypocrisy and violence, and possesses to the highest degree forbearance, straightforwardness and the rest without any conscious effort; he has no more to practice them as a discipline." (*Naiskarmyasiddhi*, 4. 69).

The *Sreyomarga* also says, "All that precede the acquisition of realization are means that are brought about by effort, but they are inherent in a perfect man." The aspirant will always be yearning for liberation, that is a form of desire, though exalted, but in a perfect sage no selfish desire remains.

However, a *Jivanmukta* may retain his higher ego to undertake welfare service as God's service. He is not satisfied with his individual salvation, but works for the salvation of others. He demonstrates by his life and action the reality of Brahman and the illusion of the names and forms of the relative world. Such a person alone keeps religion alive, not the scholar. This is illustrated by the lives of the greatest men of God like Buddha, Christ, Mohammed and Sri Ramakrishna.

Fearlessness: A *Jivanmukta* is free from fear. Most people live like hunted criminals. They are so attached to their life, always afraid to lose possessions, position, name and fame, and above all, fear of death. They cannot sit in peace or move about fearlessly. But a *Jivanmukta* has no attachment to anything and so has no fear of losing anything. He has nothing to hide and so he is fearless.

Free from Hatred and Anger: The realized sage is free from hatred (*N.B.S.*, 5). Hatred is directed toward some object or person that causes pain or injury to oneself or obstructs one's desire. It is the obstructed desire that reappears in the guise of hatred and anger. So how can a desireless sage hate anything? How can one, who perceives his own Self in the Self of every being, hate anyone? (*Isavasya Upanishad*, 6). How would a *Jivanmukta* react when someone hurts or insults him? The best illustration of that is given in the "Song of the Mendicant" in the *Bhagavatam*.

The mendicant was badly hurt and insulted by some ignorant people. And he walked on, saying to himself, "Even if thou dost think another person is causing thee happiness or misery, thou art really neither happy nor wretched, for thou art the Atman, the changeless Spirit. Thy sense of happiness or misery is due to a false identification of thy Self with the body, which alone is subject to change. Thy Self is the real Self in all. With whom should thou be angry for causing pain if accidentally thou dost bite thy tongue with thy teeth?" This is the attitude of the realized sage who has become one with all creation. Moreover, to him everything happens only by the will of the Lord, and if, therefore, he hates anyone, it will be equal to hating God Himself.

Relinquishes both Joy and Sorrow. The *Jivanmukta* does not rejoice over anything. (*N.B.S.*, 5). The *Bhagavad Gita* also affirms,

"He, who is everywhere without attachment, on meeting with
Anything good or bad neither rejoices nor hates.
His wisdom is fixed." (2. 57)

Sankaracharya also affirms in his *Vivekacudamani*:

> "He is beyond worldly fleeting pleasures and pain. He is always immersed
> in the higher joy of Supreme Bliss. How he would discard that eternal
> enjoyment and revel in fleeting things?" (Verse 522)

As Plotinus says, "they are no more two, but one; the soul is no more
conscious of the body or mind, but knows that it has what it desired,
and that it is where no deception can come, and it would not exchange
that bliss for all the heaven of heavens."

Contentment: The realized sage attains supreme contentment.
(*N.B.S.*, 4). Lord Krishna says in the *Bhagavad Gita*,

> "Ever content, steady in meditation, possessed of firm conviction,
> Self-controlled, with the mind and intellect dedicated to Me,
> My devotee, is dear to Me. (12. 14)

This contentment is not the kind of satisfaction that comes when some
desire is fulfilled; it is an absolute satisfaction arising from the absence
of desires. Will such a perfect soul lapse into inactivity? No. He is
made use of by the Lord as His instrument to carry out His inscrutable
purpose in this world.

Life of Sacrifice: A *Jivanmukta* sees the whole world as a manifestation of God; and every one of his activities will be an expression
of his devotion to God. Such a devotee's entire life becomes one
yajna (sacrifice). Having attained the Supreme Love, the realized
devotee always thinks, speaks, hears and sees the Divine Love only.
He is immersed in Divine Love. *Chandogya Upanishad* says,

> "Where one sees nothing else, hears nothing else, knows nothing else,
> That is the state of infinite bliss. . . . In that which is infinite
> Therein lies immortality." (7. 24. 1)

The *Jivanmukta* thinks of nothing else, he talks of nothing else, he
sees and hears nothing else. (*N.B.S.*, 55)

Retains individuality. Even after the realization of oneness with the Lord, the *Jivanmukta* retains his individuality as long as he lives with his physical body. Though fully conscious of his oneness with the Lord, he still loves the Lord and enjoys His sweetness and serves Him. Only love of the Absolute, eternal Truth, is the greatest. (*N.B.S.*, 81). However, we should not forget the difference in outlook that is brought about by the Realization of the Oneness with the Lord.

The Lord of his heart is a different and separate Being from himself until the sage attains realization. He now finds the Lord to be one with his own higher Self. The object of his love now is not the Personal God with an individuality of His own, but the Absolute. The sage now passes beyond all relativity of time, space and causation, beyond the three *Gunas*, beyond the three states of waking, dream and deep sleep, even beyond the subject-object relationship. He realizes his oneness with the Lord, and at the same time enjoys the sweetness of his loving relationship with Him. This experience of love is the highest experience of a devotee, superior even to *Mukhya Bhakti*.

The Behavior of a *Jivanmukta*

No uniform standard of behavior can be expected of a realized sage. Having surrendered his whole being completely to the Lord, he is always immersed in the bliss of Liberation and service of the Lord. To all external appearance, however, he may sometimes behave just like any ordinary person, discharging all the duties pertaining to his station in life, and thus setting an example to ordinary people. Often he may seem to override the accepted codes of social customs and conventions. At other times he may appear to be inactive, being immersed in the bliss of *Samadhi*, and appear dead to his surroundings like a stone. So it is very difficult to judge from his external behavior whether he is a perfect one or not. Even when the *Jivanmukta* is active externally, he is internally calm and quiet.

Sri Ramakrishna described a perfect soul in the following way:

"Sometimes he will act like a five-year old child or he may seem intoxicated or mad. Or he may be apparently inert – silent and motionless. He is not bound by any law or code, but nothing he does will be immoral and unethical. His actions are selfless without his trying to be selfless, just as a flower emits fragrance."

A realized devotee may become intoxicated with bliss. He is compared to the bee that gets intoxicated by drinking honey. Ramprasad, a devotee of the Divine Mother sang: "My mind is intoxicated because I have drunk the nectar at the blessed feet of the Mother, but to drunkards I seem to be a drunkard." The constant enjoyment of Divine Bliss makes the devotee spiritually and mentally pure and healthy.

In the *Vivekacudamani* Sankaracharya describes his experience:

> "The Ocean of Brahman is full of nectar, the joy of the Atman. The treasure I found there cannot be described in words. The mind cannot conceive of it. My mind fell like a hailstone into that vast expanse of Brahman's Ocean. Touching one drop of it, I melted away and became one with Brahman. And now, though I return to human consciousness, I abide in the joy of the Atman."

European mystics also often compare the state of Realization to a state of intoxication or madness. Plato in his *Phaedrus* calls it "saving madness". In the *Fioretti*, it is told of John of Parma how he drunk from the "chalice of the spirit of life" delivered by Christ to St. Francis. The 'trances' of Socrates, the 'union' of Plotinus, the 'vision' of Porphyry, the 'convulsions' of George Fox, and the 'illumination' of Swedenborg all are of this kind. When the devotee becomes fascinated and loses all power of action as a person dead or drunk, he loses all capacity for independent motion.

The Glory of the *Jivanmukta*

The realized devotee is the walking God upon the Earth; his body is the true temple of the Lord. Such a devotee, ever praying in the *sanctum sanctorum* of his heart for the welfare of all, moves about in society, unconscious of himself and how others consider him. (*N.B.S.*, 70).

He knows it not, but it is his presence that sanctifies the places, glorifies action and lends authority to the scriptures. The places where such saints live or have lived become holy and sanctified.

The past generations rejoice in their fulfillment, and the Divine Beings in heaven dance in ecstasy in the presence of a realized devotee living in the world. (*N.B.S.*, 71). In such a living sage all fathers of the spiritual culture, the great saints of the past, find their fulfillment. Their collective actions, their spiritual contributions, and their personal sacrifices have all been fulfilled in the realized sage, and the earth gets a savior.

Only a realized saint can save the world. All others are interested only in a world's catering to their own self-aggrandizement and enjoyment. They never have the welfare of the world at heart. Their interest in the world is like that of a butcher in his kid or a peasant in his cattle. It is the loving service of the selfless saints alone that really leads human beings to their destination. It is they who bring to the Earth the Light that dispels the gloom and darkness in the bosom of human beings

33

World Teachers and Great Devotees

WORLD TEACHERS

Introduction

Whenever evil prevails upon the Earth and righteousness (*dharma*) declines, God comes down upon earth as a Savior of humanity in response to the prayers of suffering mankind. Out of His compassion and all-consuming love for His children, God incarnates as a human being in order to show them once more how to ascend toward Divinity and to reestablish righteousness. Such Incarnations (*Avatars*) are the great prophets who have founded the great spiritual movements from time to time. They are themselves examples of the total spiritual illumination. They are worshipped by millions of followers as special manifestations of the Divine. Their love and compassion for suffering people is boundless.

The life and mission of some of the Incarnations is described below.

Sri Krishna

Sri Krishna is considered by Hindus to be the greatest of all Incarnations. He was the embodiment of Love. Right from his childhood he manifested boundless love. As a boy he lived like an ordinary cowherd boy playing with his companions. He was the adored darling of the cowherd boys and girls of Brindavan in the northern part of India. They loved him so much that they could not forget him even for a short while. When he came of age he went to Mathura where his maternal uncle, the tyrant Kamsa, was ruling. Kamsa had usurped the throne after imprisoning his own father, Ugrasena. Krishna killed Kamsa in a duel and freed Ugrasena. He restored the kingdom to Ugrasena. After this, Krishna went to Dvaraka with followers and established a New Kingdom but he himself did not become the king. Ugrasena continued to be the king at Dvaraka also. Wherever there was a rise of injustice he interfered. He enabled the Pandavas to get back their kingdom from their cousins Kauravas. This was the central story of the great epic, the *Mahabharatam*. In the great battle of Kuruksetra, Sri Krishna did not fight but only acted as Arjuna's charioteer and guide. It was in that battlefield that He gave the message of the *Bhagavad Gita* to Arjuna. He always practised the detachment in action that he preached through the *Gita*.

Though Sri Krishna lived the life of a householder, he was himself totally free and unattached to the world. But he had boundless compassion for all kinds of people–from the poorest cowherd boys and girls of Brindavan to Princes and Princesses. He was the embodiment of spiritual power and love.

> Wonderful is the teacher, Sri Krishna;
> Wonderful are His deeds.
> Even the utterance of His holy name
> Sanctifies him who speaks, and him who hears. (*Bhagavatam,* 10. 1)

The essence of Lord Krishna's teaching is:

Many are the means described for the attainment of the highest good, such as love, performance of duty, self-control, sacrifices, etc. But of all, love is the highest. Love and devotion make one forgetful of everything else. Love unites the lover with God.

Neither by Yoga, nor by philosophy, nor by deeds, nor by study, nor by austerity, nor even by renunciation of desires is God easily attained.

Only those who have pure love for God find Him easily. Blessed are the pure in heart, for unto them is given the wisdom of God.

Blessed is human birth; even the dwellers in heaven desire this birth; for true wisdom and pure love may be attained only by a human being. The purpose of this mortal life is to reach the shore of immortality. Union with God, the Soul of souls, is the end to be sought.

One who worships the Lord steadfastly with devotion soon attains purity of heart and realizes Him. When he realizes the Lord, he is free from suffering and the cycle of births and deaths. He becomes one with God.

Buddha (Sixth century BC)

Buddha is one of the Incarnations of God. He is a great lighthouse of spirituality. Six hundred years before the birth of Christ, he was born as a prince in the ancient city of Kapilavastu, Northern India. Although brought up and living in the lap of luxury, he was deeply moved by the sorrows of life and the misery of people. He renounced the world, leaving his princess and newborn son behind. After years of hard austerities, he attained to Enlightenment, with lasting peace and fulfillment. His deep compassion for the suffering humanity made him come down from that lofty height, and work among people.

Buddha's doctrine was this. Why is there misery in our life? Because we are selfish. Destroy all delusions; what is true will remain. As soon as the clouds are gone, the sun will shine. Be perfectly unselfish; be ready to give up your life even for an ant. Give up all superstition; work for your release.

Buddha's wisdom, starting from India, gradually spread, country after country, in all directions.

Buddha was born for the good of people. His boldness, fearlessness and tremendous love for people deserve veneration. He was not only great in life, but was great in death. He ate the food offered to him by a member of a tribal community that eats indiscriminately. He told his disciples: "Do not eat this food, but I cannot refuse it. Go to the man and tell him he has done me one of the greatest services; he has released me from the body."

Mahavira Vardhamana (599 BC–527 BC)

He was the last Trithankara of the Jains and a great teacher of Jainism. He was a contemporary of Buddha. He was the son of Siddhartha, a chief of warrior clan. He was born at Kundapura near Vaisali in North Bihar, India, He is said to have renounced the world at the age of thirty in order seek salvation and to have wandered for twenty years. At the age of forty-two he attained Enlightenment and became a *jina* (conqueror).

Mahavira taught his doctrine for about thirty years, founding a disciplined order of naked monks. He died at the age of seventy two at a village near Patna.

Jesus Christ

Jesus Christ is another Incarnation of God. He was born into the family of a poor carpenter, Joseph, and his wife Mary. We know very little about his boyhood. At the age of thirty he felt an inner call to go about Palestine preaching a new way of salvation for the people. He stressed the need for total purity in life and intense love for God. Jesus taught a dynamic spiritual life, a life of intense devotion and effort for self-purification. Some of his teachings are:

> "Not everyone that saith unto me, Lord, Lord, shall enter into the Kingdom of Heaven; but he that doeth the will of my Father which is in Heaven."
>
> "Be ye therefore perfect, even as your Father which in Heaven is perfect."
>
> "And ye shall know the Truth, and Truth shall make you free."

Christ's unique message is: "Love thy neighbor as thyself." It is a love based on the soul's kinship with God. His heart was full of compassion for the poor and suffering. He called to the people: "Come unto me, all ye that labor and are heavy laden and I shall give you rest." He was a true monk unattached to his kith and kin, looking upon all with equality of vision. One day when he was seated in the midst of a crowd, somebody came to him with the message that his mother and brothers were waiting to see him. But Jesus said, pointing to the devotees: "These are my mother and brothers." Then he made a remarkable statement: "Whosoever shall do the will of God the same is my brother, and my sister, and mother." The central idea of Christ's teaching is God-realization, the realization of the Kingdom of God within. "Blessed are the pure in heart, for they shall see God." This teaching is a constant reminder to people immersed in worldliness and sensuality, about what true religion is.

Mohammed (*c.* AD 570-632)

He was the prophet and founder of Islamic religion. He was born in Mecca. He was the only child of his parents: Abd Allah and Aminah. His father died before his birth and his mother also died when he was six. At the age of twenty-five he married a wealthy widow Khadijah. Mohammed engaged in trading until he received the call to be the Messenger of God. In later years after the death of Khadijah, he gradually increased the number of his wives to nine.

He is said to have spent a month each year in a cave near Mecca meditating. With the approach of his fortieth birthday, he had some startling experiences from which he emerged as the "Messenger of God." He denounced the ancient paganism and its supporters. Astounded by his audacity, the Meccans turned against him with hate and fury. Mohammed fled to Medina, the city of his maternal uncle, in 622. Many Arab tribes became Muslims and Mohammed became the ruler of Medina. He launched several expeditions against his enemies of Mecca.

He made the religion of Islam the basis of Arab unity. His messages were collected to form the Quran, the holy book of Islam. Allah is the sole Creator and sustainer of the Universe.

The doctrine of service in alleviating the suffering and helping the needy is an integral part of Islamic teaching. Hoarding of wealth without helping the needy is punishable in the hereafter. The practice of usury is forbidden.

Guru Nanak (AD 1469-1538)

Guru Nanak was the founder of Sikhism. It is a religion of synthesis of Hinduism and Islam. Nanak was born in a village near Lahore in Punjab. From early childhood Nanak was of deep religious temperament and practised meditation. At school he studied Hindi, his native dialect, and also the Persian language.

Noticing his otherworldly nature, his father made him look after farming. But Nanak's whole heart was set on spiritual matter and he devoted more time in contemplation.

When he came of age, he was married and he had two children, but he paid little attention to household duties and spent his time in the woods and in solitude one of his relatives got him a job under the Muslim Governor of the district. He and his Muslim, companion along with a few others, used to spend their spare time in singing devotional songs.

Finally he renounced the world and distributed his wealth among the poor. Accompanied by his Muslim friend he wandered all over India visiting holy places of Hindus, Buddhists and Muslims. He also visited Mecca and Medina in Arabia. He taught the people the supremacy of spirit over matter.

After returning to the Punjab, Nanak built a charitable home. His fame spread in the neighborhood and people flocked to attend his group prayers. He preached "God is one. His name is *Sat* (Truth). All are the children of the same God." He called Him Hari. He gave great importance to the repetition of the Divine Name and purity of heart. He condemned caste restrictions and exhorted people to lead a simple life.

Nanak lived and died as a great Hindu saint. After his death, his followers started a new religion known as Sikhism.

Sri Ramakrishna (AD 1836-1886)

Sri Ramakrishna was the God-man of the last century. He was born in Kamarpukur, a remote village in Bengal (India). Even in childhood he used to get mystic experiences. Later on he became a priest in the Kali temple at Dakshineswar, Calcutta. That was the beginning of a remarkable series of searching and experiments in spiritual experience, lasting for twelve years. Through intense devotion and meditation he got the vision of Kali, the Divine Mother. Even while retaining consciousness of the outer world he would see the Divine Mother.

He then tried to realize God following the various paths of Hinduism. He attained success in all those paths in a remarkably short time and experienced Divine Bliss in various ways. Later he practiced the Advaita Vedantic way and attained *Nirvikalpa Samadhi* in a single day. Not being contented with all this, he followed the paths of Islam and Christianity and was blessed with the vision of Mohammed and Jesus Christ respectively. In both cases, the figure merged in Sri Ramakrishna and he went into communion with Brahman, the same impersonal experience of God that he had attained through Hindu disciplines. He then came to the conclusion, based on direct experience, that all religious paths lead to the same goal, namely, God-realization.

The rest of his life was spent in an unbroken communion with God. He saw the supreme Spirit manifest in all beings. Utterly free from material taints, a blazing example of renunciation and knowledge, he lived like a child of God loving all without any distinctions of caste, creed or social status. He answered the questions of those who visited him day after day. A number of people and young intellectuals—Hindus, Brahmos, Christians, and Muslims–flocked to him and were elevated by his spiritual power. He trained a band of monastic disciples, including the world famous Swami Vivekananda. They established the Sri Ramakrishna Mission in India and Vedantic Centers in America and Western countries that spread his message far and wide.

Sri Ramakrishna was an Incarnation of God, especially commissioned to achieve a great spiritual mission. He was the first prophet in the historical era to demonstrate the unity of all religions by direct experience. Sri Ramakrishna re-established the ideal of direct Super-conscious Realization of God. He stressed the need for moral purity in life and harmony of religions. In the words of Romain Rolland, "Ramakrishna is the consummation of the two thousand years' spiritual life of three hundred million people."

The Gospel of Sri Ramakrishna carries Sri Ramakrishna's conversations with his disciples, devotees and visitors, recorded during the period from March 1882 to April 24, 1886. (See *The Life of Sri Ramakrishna*, Calcutta: Advaita Ashrama, 1977)

Sri Sarada Devi (AD 1853-1920)

Sri Sarada Devi was the pure consort of Sri Ramakrishna. She ranks among the world Teachers. In the religious history of India she is a unique personality. She was the visible representation of the Blissful Mother. Her marriage with Sri Ramakrishna was for the purpose of fulfillment of a great spiritual Mission, and not to live as normal husband and wife. All through their living together Sri Ramakrishna worshipped her and considered her as the Divine Mother Kali Herself. She adored Sri Ramakrishna as a Divine Incarnation. Yet she tended him all the time as a mother tended her child.

After Sarada Devi attained the highest illumination, she became an embodiment of the Divine Mother's power. She manifested that power as the mother and teacher of hundreds of people who flocked to her and to whom she granted protection, solace, peace, purity, and spiritual awakening according to their needs.

Her early days were spent in household chores, helping her parents and looking after her brothers and sisters. At Dakshineswar she served her Divine husband with one-pointed devotion and also acted as a mother to his young disciples. After the Master passed away she continued to work for the growth and integrity of the Ramakrishna Order and Mission.

In the words of Swami Vivekananda, "Without *Sakti* (Energy) there is no regeneration for the world. . . . Mother had been born to revive that wonderful *Sakti* in India; and making her the nucleus, once more will Gargis and Maitreyis be born into the world." Never before had the world seen such a great manifestation of Divine Mother as in Sri Sarada Devi, the Holy Mother. She stands as a great ideal for the women of the world, and as the ever-compassionate, ever-pardoning, all-loving mother for thousands of weary souls. Some of her teachings are:

"The only ornament of a woman is her modesty."
"Never look for faults of others. Look at your own faults. Learn to make the whole world your own. No one is alien to you, all humanity is indeed one family only."
"May all be blessed here and hereafter."

DEVOTIONAL SAINTS OF HINDUISM

Every religion in the world has produced a large number of realized saints. Deriving their inspiration from the prophets, these holy men and women tried to carry on the message of spiritual life to the doors of the common people. It is the silent work of these saints, through the centuries, that has enabled humanity to maintain its culture without falling down to the level of brutes.

Sri Sankaracharya (AD 788-820)

Sri Sankaracharya was one of the greatest philosophers of India, and an exponent of Non-dualistic Vedanta. He was born in the village of Kaladi on the West Coast of Southern India. After completing the study of the Vedas, he renounced the world at an early age in quest of Truth and was initiated into monastic life by the great ascetic Govindapada.

After attaining the highest spiritual illumination he returned to the relative plane in order to revitalize the *Sanatana Dharma*, the eternal Religion of the Hindus; and with that end in view he wrote commentaries on the *Upanishads, Bhagavad Gita* and *Brahma Sutras*. He also wrote several independent works: *Annapurana Stotram, Aparoksanubhuti, Atmabodha* (Self-knowledge), *Dakshinamurti Stotram, Nirvana Satkam, Sivamansa Puja, Vivekacudamani* and others.

He became the personification of the wisdom of the *Vedas*. He traveled the length and breadth of India, preaching the Divinity of the soul and the oneness of existence. Though a hard-core philosopher, Sankara was full of the tender love and devotion for the great gods and goddesses of Hinduism in whose praise he wrote a number of hymns.

In the midst of his incessant activity he did not, however, forget his mother. He had a tender heart. When one day he intuitively came to know that his mother was on her deathbed, he hurried to his native village. He enabled her to get the vision of Vishnu, the aspect of the Divine she had worshipped.

Before his death at the age of thirty-two at Kedarnath in the Himalayas, Sankara established monastic centers in the Four Corners of India.

He met opponents from other schools in open debates, refuted their views, and re-established the supremacy of Non-dualistic Vedanta. In him one finds the unusual combinations of philosopher and poet, savant and saint, mystic and religious reformer, debater and passionate lover of God. He stressed the importance of performing actions as an instrument in the hands of God, and of surrendering to God the fruits of actions.

Alvars

Alvars are a group of Tamil saints who flourished between the seventh and ninth centuries AD in the Southern part of India. They were all worshippers of Vishnu. They were great mystics and recorded their experiences in inspiring poems.

Altogether, the were twelve Alvars. They held the Supreme Spirit to be dearer than the dearest and that is why we find in them ecstatic love. These great saints approached the Lord with different attitudes. They loved the Lord for the sake of love without aspiring to anything, not even Liberation.

The greatest among the Alvars was Nammalvar. He composed four poems that are often called the four Tamil Vedas. In one of them he speaks of the Lord as the all-pervading Spiritual Principle who also assumes the most beautiful human form for the sake of the devotees.

Andal

Andal was a daughter of Periyalvar, one of the Alvars. She is revered as one of the greatest women saints of India. From her childhood she had no other thought except for Krishna. She looked upon herself as the bride of the Lord. Tradition holds that the Lord received her in the shrine and she merged into the Lord's Divine body.

Andal was a gifted poet. Her poem *Tiruppavai* is sung all over Southern India during the month of Margali. She is one of the best examples of bridal mysticism in India. The purified soul longs for the eternal presence of the Lord who is the Soul of all souls.

Ramanuja (AD 1017-1137)

Ramanuja was a great philosopher of Southern India. He was the first great teacher to place Bhakti on a philosophical footing. He first studied Hindu scriptures under a famous Advaitic scholar. His devotional mind could not, however, accept the Advaitic doctrine, and he developed his own ideas in which he tried to synthesize devotion and knowledge. He wrote commentaries on the *Brahma Sutras* and *Bhagavad Gita*. His philosophy known as *Visistadvaita* (Qualified Non-dualism) holds that God, the souls and the universe together form one Reality. God is the all-pervading supreme Spirit. The universe comes out of Him and returns to Him in cycles. Each cycle lasts several million years. All souls are likewise totally dependent on Him. In the spiritual path, Self-realization comes first and then God-realization. According to Ramanuja, this is possible only through God's Grace. He thus gave Bhakti supreme importance as a spiritual discipline.

Ramanuja accepted a so-called Untouchable as one of his Gurus. His teacher initiated him into the *mantra* of 'Narayana' and instructed him not to reveal it to anyone. The punishment of disobeying one's Guru was going to hell. But Ramanuja climbed the tower of the temple and openly revealed the *mantra* to the large number of people who had gathered there. On being questioned by his teacher about the propriety of his conduct,

Ramanuja said that if by his going to hell so many people could be saved, he would prefer that to his salvation.

Ramanuja traveled far and wide in India in order to spread his views. His philosophy has influenced all subsequent religious movements in India. The Bhakti movement in Northern India is traceable to his influence.

Nayanmars

Nayanmars are realized Tamil saints who worshipped the Lord Siva. They form a group of sixty-three saints, and their images are found in all the important Siva temples in Tamil Nadu, India. They belonged to various castes. The four most important among them, called the great religious teachers, lived between the AD seventh and ninth centuries. They were Appar, Jnanasambandhar, Sundramurti and Manikkavacakar. The hymns of the first three are together called *Devaram* and those of the last saint are called *Tiruvacakam.* In their devotion to the Lord, they followed the path of servant, the path of son, the path of friend, and the path of knowledge respectively. The lives of these great saints are full of miracles.

Appar

He belonged to a farming community. He embraced Jainism in early age, but was re-converted into Saivisim by the influence of his elder sister. He then spent his life moving from place to place singing the glories of Lord Siva and doing humble work in temple premises. His poems reveal a mind of high order. In one of the poems he sings,

> As fire in wood, as ghee in milk,
> The luminous One lies hid within.
> First fix the churning rod of Love,
> Pass round the cord of intelligence,
> Then churn, and God will bless thy sight.

Jnanasambandar came from a Brahmin family. When he was a small child, Goddess Parvati came down to the earth and fed him as a mother. After that he blossomed into a child prodigy and a great saint. He died at the age of sixteen.

Sundaramurti was the third great saint. At the time fixed for his marriage Lord Siva Himself came in the guise of an old man and claimed him to be his bond-slave. He thereafter wandered all over the Southern India praising Lord Siva.

Manikkavacakar was the fourth great saint. In his youth he became the chief minister of the Pandya King. The book of his hymns called *Tiruvacakam* is one of the most popular devotional books in Southern India. There is a Tamil saying that declares that nothing can melt the heart of the man whose heart does not melt on hearing the hymns of this saint. He preached Love for God as the means of attaining Him. But his Love of God is tempered with knowledge. In *Tiruvacakam* he says,

> Beyond created time and space art Thou without beginning and end.
> Yet Thou dost create many worlds, preserve, destroy and procreate.
> And through manifold births, Thou leadest me
> By Thy Grace to dedicated service.

This great book is full of deep insights into the mystery of creation, God's real nature, and the soul's experience of the Divine. But everything is blended with intense love. The book also describes the various stages through which the soul passes until it is united with the supreme Spirit.

Basavaraja (AD twelfth century)

Basavaraja was a great devotee of Siva and a great reformer. He propounded and popularized the faith of Veerasaivism in Karnataka, Southern India. Basava became the minister of the local king. He was an able and enlightened administrator. He executed many projects for the benefit of the people. He brought about many religious and social reforms. Under his leadership, the Veerasaiva religion was reformed. It accepted into its fold persons from all walks of life without distinction of caste, creed or sex. The untouchable, the barber and the brahmin all lived together and accepted even inter-caste marriages. Basava prohibited animal sacrifices, preaching about the sanctity of all life, like Buddha had done in ancient times.

Basava was a poet and composed many poems. He was an embodiment of love and compassion. In the evenings, after day's work was over, devotees used to assemble at Basava's house and used to sing songs on Lord Siva and to partake food together at his house.

Basava's teachings are: God is an absolute Reality. He is pervading the universe like fire hidden in water and the fragrance in flower. The Lord is to be realized here and now. It is only through genuine yearning the Lord responds.

Jnanesvar (AD 1275-96)

Jnanesvar was a great saint of Maharashtra, Western India. He and his eldest brother were trained in the doctrine of Natha tradition with Vedanta. The brothers and their sister wandered all over the country singing the glories of God.

Jnanesvar composed the *Jnanesvari*, one of the greatest mystical works, the *Amrtanu bhava* and many *abhangs* (lyrics) in Marathi language. Though *Jnanesvari* is a commentary on the *Gita*, it actually expounds his own synthetic philosophy and personal spiritual experience. It is the confluence of the four great paths–Karma Yoga,, Raja Yoga, Bhakti Yoga, and Jnana Yoga. According to Jnanesvar, God is both immanent and transcendent. He gives the example of the ocean and the waves. God is the ocean and the souls and the universe are waves and bubbles.

Purandaradasa (AD 1480-1564)

Purandaradasa is the father of the Carnatic or Southern Indian system of music. Originally a wealthy man well known for his miserliness, he later on gave up all his wealth and lived the life of a humble devotee of Lord Vittala, the famous Deity of Pandharpur in Maharastra. The lord tested Purandaradasa in several ways to purify him and thus turned him into a fully illumined soul. His delightful songs can be heard all over Southern India. He popularized the Path of Devotion.

He spent his last days in an Ashram built specially for him by the Vijayanagar Emperor. Some of his teachings are:

There could be no religion without morality. Observance of religious rites without morality and devotion is futile.

Deep faith in the Lord is the first step for a devotee. With sincere and strong faith in the Lord and His mercy, the devotee must pray to Him, and the prayer will be instantly answered.

What is the use of dipping one's body in the water, if there is no devotion at heart? Bathing in different rivers cannot purify the mind. Giving charity to the poor and needy is the real bath.

For God-realization one must first build up unwavering faith in the mercy of Lord and should learn to see Him in all living beings.

Assiduously serve your fellow beings and thus purify yourself. Then the Lord Himself will come to you.

Mirabai (AD 1547-1614)

Mirabai is the most famous of the women saints of India. She can be ranked among the foremost of the mystics of the world. She was a Rajput princess. Even from her childhood she was passionately attached to Sri Krishna in the form of Gridhara Gopala. When she grew up she was married to the prince of Chittor. But shortly after the marriage, both her husband and her father-in-law died. Her husband's brother became the King. Mira spent her time in prayer and singing and dancing before the image of Krishna installed in her private shrine. This attracted devotees and holy men and she was happy in their company. This was considered a disgrace to the royal household, and the king tried several methods to persuade and persecute her so that she may give up her spiritual 'madness.'

Finally, unable to bear the sufferings she left Chittor and went on a pilgrimage. After visiting Brindavan, Mathura and other places where she met several well-known saints, she finally settled at Dvaraka.

The Lord made Mira an instrument for the propagation of the message of Bhakti. The Bhakti that she taught was of the highest kind – boundless love for and total surrender to the Lord. Only a few rare souls can follow her example. About her own practice she speaks: "Tears rear the creeper of love on which, when it flowers, the Lord comes to play as the bee." She says, "I discovered the great secret in uttering the Name and adhering to this quintessence of scriptures. I reached my Gridhar through prayers and tears." Her whole life was like a flooded river of love rushing towards the ocean of the Divine.

In Mira's songs all the modes and attitudes of Bhakti find their fullest expression. Her songs are one of India's richest heritage. The spontaneous expression of her whole-hearted and intense longing for the Divine in her songs will continue to move the hearts of devotees forever. The central theme of spiritual life, according to Mira, is to love and live for the Lord. Without pure love the Lord cannot be attained.

Kabir (AD fourteenth century)

He was a great devotee and a mystic saint who tried to achieve a synthesis of Hindu and Muslim beliefs and freely used symbols from both religions in his poems and songs. He had deepest philosophical insight combined with mystical experience.

Kabir was either a real or adopted son of Muslim parents who were weavers living in Banares. He did not have a regular schooling or training in philosophy. He was married and had a son and a daughter. His family life was not a happy one. He earned his livelihood by weaving, and composed his poems while plying the shuttle.

Kabir was attracted to the great saint Ramananda (AD 1299-1410), then living in Banares. Ramananda was the fountainhead of the religious revival in the Northern India through Bhakti, started by Ramanuja in the South. Kabir very much wanted to become a disciple of Ramananda, but was afraid that the great saint might not accept him, a Muslim, as his disciple. So he hit upon a plan. He knew that Ramananda used to go for bath in the Ganges every day early in the morning. One day he went and lay down on one of the steps leading to the river. When the saint came, he did not notice the man in the darkness, and his feet touched his body. Taken aback, Ramananda exclaimed, "Ram, Ram." Immediately Kabir stood up and with folded palms addressed the saint, "Master, you have given me the *mantra* and thereby have made me your disciple, even though I am a poor Muslim weaver." Deeply moved by Kabir's devotion and humility, Ramananda accepted him as his disciple. Since then Kabir had a God-intoxicated life. Ramananda's message of simple devotion and social equality spread all over Northern India through Kabir's songs.

He condemned the blind worship of images in temples and the superficial devotions in mosques. What everyone ought to do was to seek God in the depths of his heart. In one of his poems, he says,

> If God dwells only in the mosque, to whom belongeth the rest of the country?
> They who are called Hindus say that God dwelleth in an idol,
> I see not truth in either sect. O God, whether Allah or Ram, I live by Thy name.
> O Lord, show kindness unto me. Hari dwelleth in the South,
> Allah hath his place in the West. Search in thy heart,
> Search in the heart of hearts; there is His place and abode.

Kabir laid great stress on the constant remembrance of the Divine. His songs spread among the masses and enabled them to grasp the essentials of true spiritual life. In him the Hindus and Muslims found a common meeting point. His poems profoundly influenced the social outlook and religious beliefs of the people of the entire Northern India.

Surdas (AD 1479-1584)

Surdas was one of the great saints who popularized the Krishna cult. He spent his early years in Agra and Mathura in Northern India. He was blind, but was gifted with the inner vision by the Lord. Soul-enthralling devotional songs on Krishna used to pour out from his heart. He managed to learn Sanskrit. Sri Vallabhacharya, a great teacher who visited the village of Surdas, was deeply moved by his songs. He initiated Surdas.

After the initiation, Surdas stayed at Govardhan doing service in the Srinath temple there. His ten thousand songs are collectively known as *Sur Sagar* (The Divine Ocean). In them he lavishes all his powers of description and his intuitive knowledge in depicting the soul-stirring episodes of Sri Krishna's life. The Emperor Akbar himself is said to have come to see and listen to Surdas. His songs continue to thrill and move the hearts of listeners.

When Swami Vivekananda stayed as a guest of the Maharaja of Khatri in the palace of Jaipur, the King arranged for a music recital by a dancer. Swami Vivekananda heard the song staying in his room. The dancer was singing the song of Surdas. She had surrendered herself at the Feet of the All-purifying Lord. Swami Vivekananda was stirred so profoundly that his heart went out in compassion for the poor singer.

Surdas stressed the importance of the practice of spiritual disciplines for God-realization. He, whose heart, is burnt by the fire of true repentance, will certainly obtain the Grace of God. By constant devotion to the Lord's feet, one can get rid of all evil tendencies. Devotion and absolute faith in the power and glory of God's Name are the only means of attaining lasting peace.

Sri Chaitanya (AD 1485-1533)

Sri Chaitanya was the greatest medieval saint of Bengal, India. He is considered to be an Incarnation of Radha-Krishna by his followers. His pre-monastic name was Gauranga. He was a brilliant logician and scholar. Once he went on a pilgrimage to the Holy Gaya. The spiritual atmosphere of the temple of Vishnupada Padma and the thoughts of the transience of the earthly life brought a great change in his mind, and a deep spiritual yearning arose in his heart. He met a great Vaishnava Saint named Ishwarapuri, and requested him to give initiation. Ishwarapuri initiated Gauranga into the Vaishnava cult.

Gauranga, the great intellectual, had now turned into a humble devotee mad after Lord Krishna. Soon after his return from Gaya, he got the vision of Sri Krishna playing in His flute. Gauranga lost his heart completely to Him. As the vision faded away, he broke down with tears. Thereafter he used to fall unconscious often. Along with his comparisons he began to sing of Krishna's Name and glory in the homes as well as in the streets.

Advaita Acharya, Nityananda, a Muslim Haridas and several others became Gauranga's disciples. At the age of about 24 years, he decided to become a monk. His loving wife was sad that he would be leaving her. Realizing that she would not be able to hold him long, she gave her consent. Gauranga took *sanyasa* initiation from Keshab Bharati, who gave him the name of Sri Krishna Chaitanya.

Sri Chaitanya along with a few disciples set out for the Jagannath temple at Puri in Orissa. As he entered the shrine, he had the vivid vision of Lord Krishna in the place of the idol of Jagannath.

Chaitanya thereafter went to the South visiting all Holy places and met great Vaishnava devotees, and saints including Madhavacharya, the great exponent of the Dualistic Philosophy.

After returning from the South, Chaitanya spent the last eighteen years of his life at Puri. Many became his disciples. Among them were two great Sanskrit scholars Rupa and Sanatana who renounced their ministerial positions and joined Chaitanya.

At Chaitanya's request, Nitayananda and other followers carried on his work in Bengal with great devotion. Rupa and Sanatana settled down at Brindavan and wrote a number of works in Sanskrit placing Chaitanya's conception of Bhakti on a philosophical basis.

In 1553 Chaitanya chose to leave the body. He entered the inner shrine of the Jagannath Temple and merged himself in the Lord. His message for all of us to follow is:

Krishna's Name if uttered even once will wash off all sins
and will kindle deep devotion in one's heart.
Love of Krishna is the crown of all religions.
Neither Yoga, nor knowledge, nor piety, nor study, nor austerity,
Nor renunciation will ever captivate Lord Sri Krishna
As devotion to Him does.

Tulsidas (AD 1532-1623)

Tulsidas was a great saint of Uttar Pradesh, Northern India. His work, the *Ramayana* (the story of Lord Rama) in Hindi, is the most popular and influential book in the whole of North and Central India. For the last three hundred and fifty years it has shaped the moral and spiritual lives of millions and millions of people in India. Tulsidas immortalized himself through this immortal epic *Ramayana* in the local language. He enriched it with a fresh sweetness and with an irresistible appeal to the hearts of all classes of people.

Tulsidas was born in an obscure village in Uttar Pradesh. It is said the child uttered the name 'Rama' as soon as it was born instead of crying. Considering this an ill omen, the ignorant parents abandoned the boy and soon passed away. The maid servant tended him for five years with great love. Then she also passed away. The boy had to beg for food from door to door. He was then picked up by a saint named Naraharidas at the command of the Lord Himself. Tulsidas studied for fifteen years mastering the *Vedas*, the Epics and the Vedanta.

After completing his studies Tulsidas married Ratnavali who gave birth to a son. He was passionately fond of his wife who was a very noble woman. One day, on returning home, he found that she had left for her father's place. Yearning for her, he followed her to his father-in-law's place though uninvited. When he met her there, she was annoyed at his unbecoming attachment and said: "Great is your love for this body of mine composed of bones and flesh. Had you offered half of that love to Rama, you would have been freed from worldly trouble and attained salvation." These sharp but wise words brought a new light to Tulsidas. It awakened him to the unreality of earthly relationships and also to the reality of the Supreme Spirit manifest as Lord Rama. He renounced the world and went on pilgrimage to the holy places. Afterwards he settled down in Banaras. His whole soul was drawn to Sri Rama and longed for His vision. Through the grace of Hanuman, he was blessed with several visions of Sri Rama.

The Lord made Tulsidas an instrument for the spread of Devotion to Rama. His Divine realizations filled him with love and sympathy for others and he was eager to share with all the blessings he received.

Because of the propagation of Bhakti by Tulsidas the demon of Evil, finding no place for him, threatened to devour him. Tulsi prayed to Hanuman who appeared to him in a dream and advised him to appeal to Sri Rama. As per this advice, he wrote *Vinayapatrika* which embodies his humility and self-surrender to the Lord. He also wrote many other works in Hindi for the benefit of people. Some Sanskrit scholars did not like this. They sent two hooligans to Tulsidas' house to attack him and to destroy his books. They found there a guard in the form of a young man with divine splendor holding a bow and arrow in his hands, and keeping watch. Realizing that Tulsi was under the direct protection of Sri Rama Himself, they fell at Tulsi's feet and turned into great devotees.

Emperor Jahangir was an admirer of Tulsidas. Maharaja Mansing and other princes used to visit him and honor him greatly.

The chief spiritual practice, according to Tulsidas, is the repetition of the Divine Name. In his *Vinayapatrika* he expresses his great faith in the Divine Name:

> The various paths outlined in the Vedas for the emancipation of the soul are good. But I seek only one shelter and that is Your Name, I seek nothing else.

Tulsidas did not found a new sect. His greatest achievements were to project Sri Rama as the supreme Ideal, to infuse in the Hindus a new faith and optimism.

Tukaram (AD 1598-1649)

Tukaram was the most popular of Maharashtra saints. His life is a favorite topic for a form of musical discourse known as *Harikatha*. By his pure devotion to the Lord Vittal he captivated the hearts of people. His countless soul-filling *Abhangas* (Devotional songs) resound and echo in each home.

Tukaram was born at Dehu, a village near Poona, in the Western part of India, in a pious family of farmers. When he was a young man, a great famine raged in his region. He lost his cattle, land, one of his two wives, and also a son. The surviving wife made his life miserable by constant nagging. All those unpleasant experiences turned his mind toward God. He began to study the works of saints Jnanesvar, Namadev and Eknath, and sitting in meditation on a quiet spot on the mountain.

He was initiated in a dream by saint Babaji with the holy *mantra* "Rama Krishna Hari." After a period of practice in seclusion Tukaram emerged as a poet and preacher. This aroused the jealousy of the Brahmins of the locality; one Rameshwara especially tried to persecute him in many ways. He caused all the manuscripts of Tukaram into flooded waters of Indrayani river. Tuka was totally broken down. He sat on the banks of the river in fasting and went on saying, "Vittal Vittal Panduranga! Take away my life also." On the fifteenth day, before the astonished gaze of all and to the joy of Tuka the manuscripts of the *Abangas* surfaced on the water and glided gently to the shore. While the devotees rushed down to retrieve the manuscripts, Tuka had a blissful vision of the Lord. Later Rameshwara became the most ardent disciple of Tuka.

Tuka's *abhangas* and his name and fame spread all over. Emperor Shivaji was also drawn to him. Once Shivaji sent him rich treasures. He asked the bearers to take them back, saying that Vittal was all his treasure. Having led an intense spiritual life of total surrender to the Lord, leaving a great legacy of sweet *abhangas*, Tukaram ascended to Heaven with his body. This was witnessed by hundreds of people. His *abhangas* reveal his intense devotion and total self-surrender to God. Some of his *abhangas* are:

A beggar at Thy door pleading I stand;
Give me an alms, O Lord! Love from Thy loving hand.
Holding my hand Thou leadest me my comrade everywhere
As I go on and lean to Thee, my burden Thou dost bear.
The self within me now is dead and Thou enthroned in its stead
Yea, this I, Tuka, testify, no longer now is 'me' or 'my.'

Let us learn to live this life of total dedication and self-surrender.

Ramprasad (AD 1723-1803)

Ramprasad was one of the greatest saints who made the worship of Durga and Kali popular and gave it a higher spiritual turn. His poems were a source of great inspiration to Sri Ramakrishna.

Ramprasad was born in a village near Calcutta. At school he gained proficiency in Sanskrit, Persian and Hindi. He was married at the age of twenty-two and had four children. He was initiated by the great Tantrika adept Agamavagisa and he began to spend his time in meditation. But when his father died he had to bear the burden of the family. He, therefore, went to Calcutta and became the accountant of a rich man. But Ramprasad's mind was too busy with the thoughts about the Divine Mother to pay attention to accounts. He wrote poems about Her, as they came to him, on pages of the account books. The matter finally reached the ears of the master who examined the books. But instead of feeling angry, he was deeply moved by the poems. He told the erring accountant that he would henceforth get a monthly allowance for the support of his family and he could go home and continue his spiritual practice without any worry.

Ramprasad went back to his village and gave his heart and soul to the worship of Kali and intense meditation on Her. While engaged in practice, songs would rise spontaneously within him, and these reflected the great spiritual struggles he had to undergo during this period. He had wonderful vision of the Divine Mother and his later life was immersed in higher spiritual consciousness. In 1803 on the day after Deepavali, when the image of Kali was about to be immersed in water according to custom, Ramprasad entered the river and gave up his body while singing praises to his beloved Divine Mother Kali.

The poems of Ramprasad are sung all over the Bengali-speaking world. Their simplicity and intense spiritual fervor have made them very popular for the past two hundred years. They are a source of great inspiration to spiritual aspirants.

Tayumanavar (AD eighteenth century)

Tayumanavar was a great poet and a mystic of Tamil Nadu, Southern India. He was one of the greatest and the most popular of Saiva Saints. As a young man he became a steward in the palace of a king and gained good mastery over Tamil and Sanskrit devotional and philosophical literature. He came across a sage who was observing silence and got attracted to him.

The widowed queen offered herself and her kingdom to Tayumanavar. He turned down the offer and left the place. He went to another place, married a pious girl and lived a virtuous life. When his only child came of age, the silent sage suddenly appeared before him and reminded him that it was time for him to renounce the world.

Tayumanavar spent the rest of his life as a wandering minstrel singing his highly philosophical poems. He reconciled Vedanta and Saiva Sidddanta. According to Him, God, soul and Nature are three ultimate categories. Just as the sun energizes the living world, so God sustains and enlivens all the souls and nature. In one of his works called *Anandakkalippu* (Revel in Bliss) he describes the final realization and shows how Bhakti and Jnana become one at the highest experience.

Tyagaraja (AD 1767-1847)

Tyagararaja occupies a unique position in South Indian Carnatic Music. He was a supreme devotee who captivated and bound Rama in his heart with the strains of his melodious devotional music.

He belonged to a pious Telugu family lived at Tiruvaiyyaru (the land of five rivers) in Tamil Nadu, Southern India. He inherited the devotion and musical talent of the family. He received his initiation of 'Rama' *mantra* quite early in life from Brahmayogi. He learnt music from a great musician and also did much self-study in music and related arts.

Tyagaraja practiced *Japa* and completed the repetition of Rama's Name 96 crores of times, and attained the vision of Lord Rama. He composed 2400 songs on Rama and wrote two devotional operas: *Prahlada Vijayam* and *Nauka-charitra*. His songs are noteworthy for his devotional fervor and ethical and spiritual preaching. He pours out his heart in his songs. They stir us to tears of ecstasy. The most of the songs heard in Southern Indian musical concerts are his compositions. On January 6, 1847 Saint Tyagaraja attained *Nirvana* (salvation) leaving behind a host of disciples.

Every year on the Nirvana day Tyagaraja Aradhana Festival is celebrated at Tiruvaiyaru in adoration of this great Saint. The highlights of the Festival are performance of worship service at the *Samadhi* and singing of the Saint's songs by great musicians and instrumentalists.

One example of his songs, in which he describes his vision of the Lord, is:

> "O Lord of my life! You have come walking all the way to bless me, to fulfill the secret longing of my heart; a vision of your lotus-eyed face is the sole objective of my life. You have been so gracious as to appear before me in Your resplendent blue-hued body, with the mighty and majestic bow and arrows in your hands and accompanied by (my Mother) Sita. Oh, how blessed indeed is Tyagaraja!"

Swami Vivekananda (AD 1863-1902)

Swami Vivekananda was the greatest of Sri Ramakrishna's disciples. The Master classed him among the ever-perfect souls who are born on earth for the welfare of mankind.

His pre-monastic name was Narendranath. He was born in

Calcutta into an aristocratic family. Even as a child he practised meditation and showed great power of concentration. In his boyhood and youth, Narendra possessed great courage and presence of mind, a vivid imagination, deep power of thought, keen intelligence, an extraordinary memory, a love of truth, a passion for purity, a spirit of independence and a tender heart.

He was an expert musician and also proficient in physics, astronomy, philosophy, history and literature. He was extremely handsome but with unshakeable chastity and purity. Under the influence of college education he became an agnostic for a short period during his adolescence. He would not accept religion in mere faith; he wanted a demonstration of God. He wanted a perfect Guru to guide him. He joined the Brahmo Samaj, but he did not find a Guru who had seen God.

At the age of eighteen while studying in college, Narendra heard about Sri Ramakrishna. Narendra came to him at Dakshineswar. He found in Sri Ramakrishna's words an inner logic, a striking sincerity, and a convincing proof of his spiritual nature. In response to Narendra's question, "Sir, have you seen God?" the Master said, "Yes, I have seen God more tangibly than I see you. I have talked to God more intimately than I am talking to you." Narendra was amazed. These words he could not doubt. This was the first time he had ever heard a person saying that he had seen God.

Early in 1884, Narendra's father suddenly died of heart failure, leaving the family in a state of utmost poverty. Actually starving and barefoot, Narendra searched for a job, but without success. One day, soon after, he requested Sri Ramakrishna to pray to the Divine Mother to remove his poverty. The Master asked him to pray Her himself. When he entered the shrine, forgetting the purpose of coming, he prayed only for knowledge, love and liberation. Thrice he went into the temple, and thrice he was unable to pray for petty worldly things. Now he asked the Master himself to remove his poverty, and was assured that his family would not lack simple food and clothing. This significant experience taught Narendra the imperative necessity of worshipping a Personal God. Under the Master's Guidance, Narendra practiced intense meditation with great faith and dedication, and at the age of twenty three was blessed with *Nirvikalpa Samadhi,* the highest spiritual experience.

After the Master passed away, and at his direction, Narendra organized the young disciples of Sri Ramakrishna into a Monastic Order and set out on a journey across the sea to America. He delivered an inspiring talk at the Parliament of Religions held at Chicago in 1893. He captivated the audience with his historical message of Vedanta. After four years of hectic preaching service in America and England and all over the world, he returned to his motherland. He received a hero's welcome. He lectured at several places from Colombo to Kashmir, rousing the sleeping nation to the glories of its ancient heritage, and to the poverty and backwardness of the masses of modern India. His great heart bled at the sufferings of the poor and the ignorant. In India he stressed social service very much, and founded the Ramakrishna Mission with this end in view.

Swami Vivekananda was a mighty spiritual personality, and his compassion for humanity was of a high order; it was based on actual spiritual kinship with the people. He saw the Self in all beings, and service of humanity was for him worship of God. His life and message are a great force shaping the destinies of millions of people all over the world. *The Complete works of Swami Vivekananda* have been published in several volumes by Advaita Ashrama, Calcutta.

Conclusion

Every part of India has produced many saints of great spiritual eminence. Through their lives, teachings, and songs, and through their organized followers, the great saints have made Hinduism a living faith adapted to the changing times. Whenever and wherever the nation had to face the challenges of atheism or foreign aggression, saints arose and gave new directions to the spiritual energies of the Hindu race. Thus Hindu spirituality has been a dynamic force though branched into a number of sects, doctrines and practices. Beneath all this diversity there has always run the basic continuity of Hindu culture. The underlying current of this culture is man's quest for freedom from sorrow and ignorance; that total freedom can be attained through direct Realization of God; and God can be realized through diverse paths. The attainment of God-realization is the goal of life.

CHRISTIAN MYSTICS

St. Paul

Christianity also has produced great mystics. Apart from the Apostles of Christ, of whom we know very little, the earliest mystic was St. Paul. As a young man he is said to have been violently opposed to Christianity. Then one day on his way to Damascus he had a wonderful spiritual experience. He saw a brilliant light. It so overpowered him that he became blind for three days, and he heard the voice of Jesus Christ. This converted him to Christianity; and he spent the rest of his whole life preaching the gospel of love for man. His whole life was centered in God in whom "we live, move and have our being." He said, "I live, yet not I but Christ liveth in me." He traveled far and wide and tried to help people to lead a God-centered life of purity and service.

St. Francis of Assisi (AD 1181-1226)

St. Francis was one of the greatest saints of Christianity. He was born the son of a rich merchant and spent his youth in merry making with his friends. But, after a protracted illness, he was converted to a spiritual life and renounced the world. One day, while praying in a dilapidated church, he heard a Divine Voice commanding him to repair that house of God. Single-handed he collected building materials and carried them all on his head. Humorously he used to call his body, "Brother donkey." He was totally detached. Filled with peace and great joy, he used to sing and dance.

He loved birds, animals, plants and even inanimate things and spoke of them as "Brother Wolf," "Brother Sun," "Sister Moon," etc. He had great respect for the written words, considering all words as come from the mouth of God.

He established the great Franciscan Order of Monks who lead a life of utter poverty like the wandering monks of India. St. Francis had great influence over the minds of people and hundreds became his followers. His life was entirely consecrated to the Divine. He had no fear of death and welcomed it as "Sister Death."

St. Thomas Aquinas (AD 1227-74)

St. Thomas Aquinas has had a profound influence on the Catholic Church. He was born into an aristocratic family . But, unlike his relations he had a passionate love for purity and spiritual life. At the age of seventeen he tried to become a monk. His enraged brothers imprisoned him in a solitary room and also tried to tempt him in many ways. He not only overcame all the temptations but also spent his time absorbed in prayer and study. At last he was set free and he joined the Order of Dominicans. He soon became the foremost theologian of the day and his book *Summa Theologica* is the most authoritative book on Roman Catholic theology. While about to complete the book he had a wonderful spiritual experience and henceforth refused to write. He left his masterpiece unfinished. When he was questioned about it he replied: "Such secrets have been revealed to me that all I have written until now appears to be of little value as useless as straw."

St. Ignatius of Loyola (AD 1491-1556)

St. Ignatius Loyola was born of a noble family. He entered the army and distinguished himself by his great valor. He was wounded in battle against Francis I, king of France. During convalescence he read about the Saints and this inspired him to leave the worldly life and devote himself to God.

He spent ten months in solitude at the monastery of Monserrat during which time he composed his book *Spiritual Exercises*. He then made a pilgrimage to the Holy Land and on his return founded the Society of Jesus also known as the Jesuits. His Order became one of the most politically powerful and wealthy Orders of the Catholic Church. The Society received the official approval of Pope Paul in 1540.

St. Theresa of Avila (AD 1515-1582)

St. Theresa was born in Avila, Spain. Her mother died when she was at twelve and her father placed her in a convent of Augustinian nuns. When she returned home she was determined to enter the religious way of life. She became a nun in the Carmelite Convent of the Incarnation near Avila and made her profession in 1534.

After many years of study she became inspired by the Spirit and acting under the direction of enlightened men such as St. Peter of Alcantara she began the task of reforming her Order and restoring the primitive observance.

Assisted by St. John of the Cross she established the reform of the Discalced Carmelites for both the Sister and the Monks of the Order. By the time of her death 32 monasteries and 17 convents had been established. St. Theresa wrote several books on mystical theology including *Way of Perfection, Interior Castles* and *Spiritual Testimonies.*

St. John of the Cross (AD 1542-91)

John Yepez was born at Fontiberos in Old Castile, Spain. Form his early childhood he had a marked devotion to the Blessed Virgin. After studying in a Jesuit college he entered the monastic Order of Carmelites. In the monastery at Medina he practiced great austerities and was ordained as a priest in 1567.

Shortly thereafter he met with St. Theresa of Avila. She interested him in reforming their Order.

He became the superior of the new Order. But in the midst of all that work his heart was always with God. He is always remembered as one of the great mystical contemplatives of the Catholic Church. Among his books the most famous is *Dark Night of the Soul.*

Brother Lawrence (AD 1611-91)

Brother Lawrence was born of poor parents. At the age of eighteen he underwent a conversion when he saw a tree stripped of the leaves in winter. He began to think about the deep mystery and power of life that would make the tree produce fresh leaves and flowers after some time. This thought weaned him away from worldly life and made him seek the Divine, the source of all life and Consciousness.

When he grew up he was for a time in the army and in the service of a nobleman, and finally became a cook in a Carmilite monastery in Paris. He worked in that humble capacity for thirty years with his soul enwrapped in God. By the constant practice of God even while engaged in the ordinary duties of kitchen, he attained a high degree of spiritual enlightenment. He soon became well known for his holiness. Even noble men and high dignitaries sought his advice. His little book, *"The Practice of the Presence of God"* contains the conversations and letters of Brother Lawrence.

About his state of mind he said:

> The time of business does not with me differ from the time of prayer; and in the noise and clatter of kitchen, when several persons are at the same time calling for different things, I possess God in as great tranquility as if I were upon my knees at the Blessed sacrament.

(Source: Hugo Hoever, *The Lives of Saints,* Catholic Book Publishing, 1977)

SUFI MYSTICS

In general, Islam is anti-monastic and anti-mystical. Yet it produced a number of great saints during the Middle Ages. Most of them were mystics. Islam mystics are called Sufis.

Rabia of Basra (AD 717-801)

Rabia was one of the earliest Saints. Orphaned at an early age, she was kidnapped and sold as a slave girl by a wicked man. Her new master was equally cruel, but one night he saw a strange light surrounding the poor girl, and getting frightened, set her free. She spent some time in the solitude of a desert and then lived like a hermit in Basra in Iraq. Her spiritual experiences put her beyond the dualities of this world. When she was asked whether she hated Satan, she replied: "No. My love for God leaves no room in my heart for hating anybody." Her most famous prayer is: "O my Lord. If I worship Thee for fear of hell, burn me in hell; and if I worship Thee for the hope of Paradise, exclude me from there; but if I worship Thee for Thine own sake, withhold not from me Thine eternal beauty."

Mansur Al-Hallaj (AD 858-922)

He is the greatest of Sufi mystics. He is also simply known as Hallaj. He was born in Southern Persia. Even in his boyhood he had a mystical temperament and sought guidance from several Sufi saints. When he grew up he went to Baghdad and became the disciple of Junayd, a well-known Master of that time. After spending some years in seclusion in Mecca, Hallaj traveled to India by sea. At that time India had not come under Muslim rule. He was one of the earliest Sufi mystics to visit India. He had lengthy discussions with Hindu mystics. On return to Baghdad he started preaching. The most famous and fundamental of his doctrines was *Ana'l-hagg* (I am the Reality) which is similar to the Vedantic dictum, "I am Brahman." The outraged Muslim theologians got him arrested and executed in a barbarous manner. But his ideas about the Divinity of soul and the possibility of attaining oneness with the Supreme Spirit influenced Islamic mysticism for several centuries.

Jalalud-Din Rumi (AD 1207-73)

Jalal was a great Islamic mystic and poet. He was born at Balk in Eastern Persia. When he was a small boy his father incurred the displeasure of the ruling sovereign and had to leave his native place with his family. After long wanderings the family settled in Qwoniya in Turkey. Jalal was educated in the Arabic universities and became famous as a great scholar. But his contact with a wandering dervish brought about a sudden conversion in his life. Thenceforth he spent a long time in contemplation. In memory of his beloved teacher, he founded a new religious order called the Maulavis who practiced a kind of whirling dance to induce spiritual ecstasy. He wrote several books including the *Mathnavi* considered one of the greatest mystical poems of all time.

Jalal looked upon God as the only Reality and the phenomenal world as a shadow of His. He believed in the pre-existence of the soul which according to him passes through an ascending series of stages of corporeal existence–mineral, vegetables, animal, human, and angelic–before it finally unites itself with God. His allegory about a man knocking at the door of his beloved is famous. The voice from inside asks, "Who is there?" and when the lover answers, "It is I," the door is not opened. Later on when he comes again and knocks, and in response to the usual question, answers, "It is Thou," the door opens. The meaning of the allegory is that as long as the ego consciousness persists, full union between the soul and God is not possible.

Jalal had a large number of disciples. His writings exercised considerable influence on the development of Sufism.

Conclusion

The study of the lives of saints belonging to different religions leads to the same conclusion. They speak the same truth though in different languages. im different names and forms. God is the Soul of all souls, the source of all bliss and peace. Our task is to follow in the footprints that these great men and women have left on the sands of time and attain supreme Bliss, Enlightenment and Freedom.

Glossary

Abhinivesha. Clinging to life

Abhyasa. Practice

Adharma. Unrighteousness

Adrishta. Unseen potency

Advaita. Non-dual

Agnihotra. Fire sacrifice

Aham. I; ego

Aham Brahamasami. I am Brahman

Ahamhara. Egoism

Ahimsa. Non-hurting; non-violence

Ajapa Japa. The mantra 'soham' produced by breath naturally

Ajna. Psychic center in the eyebrow

Akasa. Space; ether

Akhanda. Indivisible; unbroken

Anahat. Psychic center at the heart area

Anahata sounds. Mystic sounds heard at Anahata center

Ananya Bhakti. One-pointed devotion

Anavasada. Cheerfulness

Apana. The down-going breath

Arati. Waving of light before the deity

Archana. Offering flowers in worship

Arta. One who is distressed

Artharthi. Seeker of wealth

Asana. Posture

Ashram. Monastery; spiritual center

Asmita. Egoity

Asuras. Demons

Atma jnana. Knowledge of the Self

Atman. The Self

Atmanivedana. Self surrender

Avadhuta. Naked ascetic

Avatara. Incarnation of God

Avidya. Ignorance

Bhagavan. The Lord

Bhajan. Singing the names of the Lord

Bhakta. A devotee of God

Bhakti. Devotion to the Lord

Brahmacharya. Celibacy

Brahman. The Absolute

Brindavan. Indian city associated with the life of Lord Krishna

Buddhi. Intellect

Chaitanya. Pure consciousness

Chakra. Psychic center in the body

Dacoit. Robber

Dama. Sense control

Dana. Charity

Darshan. Vision

Dasya. Servant

Daya. Compassion

Deva. A god

Devi. Divine Mother

Dharma. Righteousness

Fakir. Muslim holy man

Gauna. Secondary devotion

Gayatri. Hindu mantra

Gunas. Modes of nature

Guru. Preceptor

Hari. A name of Lord Krishna

Hiranyagarbha. Cosmic intelligence

Indra. Chief of celestials

Ishta devata. One's chosen deity

Ishvara. Personal God

Isvarapranidhan. Devotion to the Lord

Japa. Repetition of a mantra

Jijnasu. Spiritual aspirant

Jiva. Individual soul

Jivanmukta. Liberated while alive

Jivatma. Individual spirit or Self

Jnana. Wisdom of the Reality

Jnani. One who attained wisdom

Kaivalya. Liberation

Kalyana. Wishing well to all

Kama. Lust

Karana sarira. Causal body

Karma. Action

Kirtan. Singing the Lord's glory

Kosha. A sheath enclosing the soul.

Kriya. Action

Kundalini. The primordial cosmic energy

Lila. Divine play

Lingam. Symbol of Lord Siva

Linga sarira. Subtle body

Madhurya bhava. Lover's attitude

Mahabhava. The highest state of devotion

Maha vakyas. Upanishadic declarations

Maheshwara. Great Lord the crown of the head

Mala. A rosary of 108 beads

Manana. Constant reflection

Manasic. Mental

Manipura. Psychic center located at the navel area

Mantra. Sacred syllable or word

Mantra diksha. Initiation into a mantra

Maya. Lord's illusory power

Moha. Delusion

Moksha. Liberation

Mukhya. Primary bhakti

Mumukshutva. Intense longing

Nada. Mystic sound heard by yogis

Neti neti. Not this, not this

Nididhyasana. Deep meditation

Nirguna. Without attributes

Nirvikalpa. Without mental modifications

Nishkama bhakti. Pure devotion

Niyama. Moral disciplines

Om. The sacred syllable
Ojas. The power produced when seminal energy is sublimated
Padaseva. Serving the Lord
Para. Supreme
Paramahamsa. The highest class of sannyasins
Paramatma. The Supreme Self
Pitri. Forefathers
Prakriti. Primordial matter
Prana. Life-force
Pranayama. Control of prana
Prasad. Blessed food
Pratima. Image of God
Pratiyahara. Withdrawal of senses
Prema. Divine Love
Prithvi. Earth own child
Puja. Worship
Purana. Book of Hindu mythology
Purusha. The supreme Being
Raga dvesha. Likes and dislikes
Raja Yoga. The royal Yoga
Rajasic. Passionate
Rishi. Sage
Sadhaka. Spiritual seeker
Sadhana. Spiritual practices
Sadhu. Pious man; monk
Saguna. With attributes
Sahasrara. Psychic center located at throat
Sakhya. The attitude of a friend
Sama. Control of mind
Samadhi. Superconscious state
Samsara. The cycle of births and deaths
Samskaras. Innate tendencies
Sannyasin. A monk
Santa bhava. Attitude of calmness
Satchitananda. Existence-knowledge-bliss
Satsang. Company of the wise
Sattva. Purity, light
Sattvic. Pure

Satyam. Truth
Shakti. Power, energy
Shatsampat. Sixfold virtues
Siddhis. Perfection; psychic power
Sloka. Verse
Smarana. Remembrance
Sraddha. Faith
Sravana. Hearing of scripture
Sushumna. A subtle channel from the base of the spine to the top of the head
Svadhisthana. Psychic center located near the root of reproductive organ
Tamas. Ignorance, inertia, darkness
Tantric. Worship of God as Divine Mother
Tapas. Austerity
Tat tvam asi. Thou art That
Tejas. Spiritual brilliancy
Upasana. Worship; sitting near
Vairagya. Dispassion
Vaishnavite. A worshipper of Lord Vishnu
Vandana. Prostration to the Lord
Varna. Color
Vasana. Subtle desire
Vatsalya. Attitude of treating God as one's own child
Prithvi. Earth
Vayu. Air; the wind
Vedas. Hindu revealed scriptures
Vedanta. The doctrine of non-dualism
Vibhuti. Sacred ash
Virat. The physical world
Vishuddha. Psychic center at the area of the throat
Viveka. Discrimination between the Real and the unreal
Viritti. Thought-wave; mental modification
Yajna. A sacrifice
Yoga. Union; Spiritual path